Resources of Water

Edited by
Prathna Thanjavur Chandrasekaran,
Muhammad Salik Javaid and Aftab Sadiq

Published in London, United Kingdom

IntechOpen

Supporting open minds since 2005

Resources of Water
http://dx.doi.org/10.5772/intechopen.77929
Edited by Prathna Thanjavur Chandrasekaran, Muhammad Salik Javaid and Aftab Sadiq

Contributors
Guenter Mueller-Czygan, Joseph Okeyo Obosi, Jesús Efren Ospina-Noreña, Carlos Gay-García, Nathaly Güiza-Villa, Cheikh Faye, Shemsu Gulta, Brook Abate, Ramzy Saeed Mahbob Naser, Driss Belghyti, Mohammed El Bakkali, Nurul Aini Amran, Siti Nor Adibah Mustapha, Manuel Pulido Fernández, Javier Lozano-Parra, Jacinto Garrido Velarde, Muhammad Salik Javaid, Prathna Thanjavur Chandrasekaran, Muhammad Zeshan Khalil

First published in London, United Kingdom, 2021 by IntechOpen
IntechOpen is the global imprint of INTECHOPEN LIMITED, registered in England and Wales, registration number: 11086078, 5 Princes Gate Court, London, SW7 2QJ, United Kingdom
Printed in Croatia

British Library Cataloguing-in-Publication Data
A catalogue record for this book is available from the British Library

Additional hard and PDF copies can be obtained from orders@intechopen.com

Resources of Water
Edited by Prathna Thanjavur Chandrasekaran, Muhammad Salik Javaid and Aftab Sadiq
p. cm.
Print ISBN 978-1-83881-912-5
Online ISBN 978-1-83881-913-2
eBook (PDF) ISBN 978-1-83881-914-9

For EU product safety concerns:
IN TECH d.o.o., Prolaz Marije Krucifikse Kozulić 3, 51000 Rijeka, Croatia, info@intechopen.com or visit our website at intechopen.com.

Meet the editors

Prathna Thanjavur Chandrasekaran, Ph.D., has been working in the field of water research for the past nine years. She has more than thirty publications in peer-reviewed international journals and an edited book to her credit. She has published more than 1700 citations and has an h-index of fifteen. Dr. Prathna has a Ph.D. in Nanotechnology and has worked with academic institutions of repute such as the Indian Institute of Science (Bangalore) and IHE Delft Institute for Water Education, the Netherlands (formerly UNESCO-IHE Institute for Water Education), as part of her postdoctoral studies. Her research focuses on optimizing adsorbent nanomaterials for the development of decentralized drinking water filters. She currently works as a chief minister`s fellow with the Department of Irrigation and Flood Control with the Delhi Government (India) and works on projects related to the cleaning of River Yamuna, the revival of water bodies, decentralized wastewater treatment, water management, and re-use.

Dr. Muhammad Salik Javaid has been working on academics, research, development, execution, management, and policy formulation assignments with the Pakistan Army Corps of Engineers and engineering universities in Pakistan and abroad for the past four decades. He has worked as Chief Consulting Engineer in the organization of Engineer-in-Chief, Director General Planning for Earthquake Reconstruction and Rehabilitation Authority, Pakistan, and a disaster risk reduction consultant with the United Nations Development Programme (UNDP). He is a graduate of Georgia Tech, Atlanta, the USA where he undertook his graduate and postgraduate studies in Dam Overtopping and Ocean Bed Hydrodynamics. He has published various research papers, and edited six books on water resources. He has conducted various courses for civil engineering students at universities in Pakistan. During his last engagement, Dr. Javaid taught Fluid Mechanics and Hydraulic Engineering at Morgan State University, Baltimore, USA.

Dr. Aftab Sadiq is a Professor of Civil Engineering at the University of Wah, Pakistan. He received a BSc (CE) from MCE, Risalpur, Pakistan, and an MS and Ph.D. from Georgia Institute of Technology, Atlanta, Georgia, USA. His field of specialization is open-channel flows and his research work focuses on clear-water scour around bridge abutments in compound channels. He has served on the faculty of the National University of Sciences and Technology (NUST), and Air University, Islamabad, Pakistan. He has taught courses in fluid mechanics, open channel flows, hydraulic engineering, and water resources engineering. He has served as a consultant to various civil engineering design, consultancy, and project implementation organizations. He has worked on various civil engineering projects related to water resources, water flows, water supply, drainage, dams, groundwater, pumping, and others.

Contents

Preface

Water is a limited natural resource indispensable for human existence and critical for socio-economic development. However, rapid urbanization and inefficient management of available natural resources have hastened the pace and impact of climate change. Floods and droughts related to climate change are a recurring phenomenon in recent years and have disrupted an already fragile ecosystem. Because the need to conserve available water resources and devise strategies for resource management is very relevant, this book deals with water resources and their management. Edited by Prathna Thanjavur Chandrasekaran, Muhammad Salik Javaid and Aftab Sadiq, the book provides relevant information on major aspects of water resources including water management strategies practiced in different parts of the world, the need for conservation, impact of climate change on water resource management, and contamination of water resources and their treatment strategies.

The first section of this book discusses aspects related to water resources and their conservation, with the first chapter introducing the reader to diverse aspects of resources of water. The chapter on water availability in Chile by Pulido Fernández compares the current and future availabilities of water for the different regions of Chile and provides relevant information on the water balance for land planners. The chapter on Smart Water by Mueller-Czygan discusses the role of software such as Industry 4.0 and KOMMUNAL 4.0 initiated by the federal government in Germany in municipal water management and in tackling future water challenges. Many small- and medium-sized communities are faced with the challenge of reliably planning for the future in terms of maintaining and expanding their infrastructure in view of the consequences of demographic change. Software such as KOMMUNAL 4.0 can play a critical role in assisting communities to manage water infrastructure digitally. The chapter by Gulta and Abate elaborates on the importance of aquifer characterization in effective groundwater management. Aquifer characterization is the first step in enabling strategies to manage it. The authors used Aquifer Test V.2.55 to characterize the aquifer system in Hawassa, Ethiopia, physically, potentially, spatially, quantitatively, and qualitatively.

The second section of the book details strategies adapted for sustainable water management in different parts of the world with the help of interesting case studies. The chapter on community water-service management in Africa by Obosi explains the critical need of community water supply with experience gained from Kenya, Tanzania, Ghana, Malawi, and Nigeria. Community-based water supply management allows communities to form their own institutions for water delivery without formal connections or partnerships with utility operators or municipal governments and can play a critical role in ensuring access to safe water in developing countries. Anthropogenic activities can hasten the process of global warming and the resulting shifts in weather defined as climate change. The chapter by Ospina-Noreña et al. explains the effect of climate change on water resources, indices, and related activities in different parts of Colombia using models. The study also reviews the influence of climate change on hydroelectric power generation and water balance in arid areas. Likewise, it outlines a possible future water supply–demand relationship, where supply is associated with a change in the water balance and demand with some crops, activities, and sectors that need water to survive. The chapter by Javaid and Khalil emphasizes the critical need to reintroduce physical hydraulic and hydrologic modeling into engineering practice with a case study of water management projects in Pakistan using physical hydraulic modeling. The chapter by Faye sheds light on water management initiatives in the city of Dakar, Senegal.

The last section of the book provides information on the evaluation of contaminated water resources and strategies to treat them. The chapter by Naser et al., for example, provides GIS and statistical evaluation of the fluoride content of a Rasyan aquifer in Yemen. Fluorine has the highest chemical reactivity among all known elements and can be present in water resources due to natural as well as anthropogenic activities. Fluoride in excess levels can cause fluorosis and is often found to be an endemic problem in locations surrounding water resources with high fluoride levels. Understanding fluoride occurrence is important in the management of fluoride-related epidemiological problems. Discharging accumulated bilge water from a ship is essential to maintain the ship's stability and safety. However, the bilge water that contains contaminants, including waste oils and oily wastes, must be treated prior to discharging to the sea. Thus, an efficient oil–water separator must be installed to separate the oil from the bilge water to minimize marine pollution. The chapter by Amran and Mustapha introduces and discusses the working mechanisms as well as advantages and disadvantages of the available oil–water separation techniques for bilge water treatment.

Although this book may not provide readers with comprehensive information on all aspects related to water resources and their conservation, it does provide constructive data and content on the current trends and advancements in the field. The book will motivate readers and scientists alike to look further and make concerted efforts towards promoting the preservation and conservation of water resources.

Prathna Thanjavur Chandrasekaran
Department of Irrigation and Flood Control,
Govt. of National Capital Territory of Delhi,
Delhi, India

Muhammad Salik Javaid
Morgan State University,
Baltimore, MD, USA

Aftab Sadiq
Wah Engineering College,
University of Wah,
Pakistan

Section 1

Water Resources Management

Chapter 1

Introductory Chapter: Resources of Water and the Need for Conservation

Prathna Thanjavur Chandrasekaran

1. Introduction

Unlike other natural resources such as wind and solar, water resources are limited and can include surface water, ground water, rain water and saltwater. Of all the water resources present on earth, only 2.5% exists as freshwater. Fresh water can exist in the form of ground water, surface water and water present in glaciers and ice. A major fraction of fresh water is present in glaciers with only a tiny fraction available as surface water.

Increasing population and urbanization has increased our water demands and place undue stress on the existing water resources. Studies indicate that by 2030 there will be a 40% deficit between the world water demand and the available water resources [1]. People in the developing world are most vulnerable to climate changes; changes in rainfall pattern as a consequence of climate change have largely affected the world in recent years. Extreme changes in the rainfall patterns have increased the frequency of floods and droughts. Exploitation of the available water resources affects all aspects of human existence in addition to disrupting the fine balance in the ecosystem. Erratic and unseasonal rainfall largely affects agricultural activity which mainly sustains the economy of most of the developing countries. Sustainable water management practices are therefore critical to ensure conservation of the existing water resources. Development of policies on smart water practices and its implementation need to go hand in hand to promote conservation of water resources. Many countries around the world have adopted different strategies to reduce the pressure on their available water resources and interesting case studies on their success stories exist.

2. Strategies for sustainable water management

Some of the strategies that can be adopted for sustainable water management are: (i) re-use of treated water for various end uses such as agriculture, horticulture, ground water recharge, aqua culture etc., (ii) revival of water bodies; (iii) reduction in non-revenue water losses. These strategies need to be executed in parallel to conserve and sustain the water resources.

3. Re-use of treated water

Treated water from the Sewage Treatment Plants (STPs) can be used to cater to horticultural needs, irrigation and partly for the daily water requirements. The city of Delhi for example has three sources of water: (1) Surface water supplied by river basins; (2) groundwater; (3) Treated water- the status of which will depend on the quality of treatment. The daily water requirement for Delhi is around 1120 MGD, of which 840 MGD is extracted from River Yamuna and 80 MGD is extracted from groundwater. Currently most of the treated water from STPs are discharged into the major drains and reach River Yamuna. As of 2019, out of 720 MGD generated as raw sewage, only 90 MGD is currently being utilized by various agencies for non-potable purposes like horticulture and in STPs. Some of the advantages of utilizing treated wastewater are:

1. It will reduce the ever-increasing gap of Potable Water Supply and Demand in Urban Cities.
2. It can bring down billing charges of fresh water which are a result of long-distance transportation, gradient and higher energy costs.
3. It can mitigate conflicts of water resource allocation between the Domestic and Agriculture/ Industry.
4. It can reduce groundwater extraction and also aid in conservation of water resources.
5. It can make water and sanitation sector sustainable.

4. Revival of existing water bodies

Water bodies are an excellent source to conserve rain water which otherwise can lead to flooding due to complete concretization of cities. Revival of existing water bodies and creation of new water bodies can have many advantages; few of which are mentioned below:

1. Additional Reserves of water can be created within the city which can be utilized in case of scarcity
2. Ground water recharge
3. Treated water coming from STPs can be utilized which currently discharge into drains
4. Rain water can be captured and stored at massive level
5. Excess water can be utilized to meet needs of the people
6. Social and Cultural connected – Community owned space – Protection of water bodies

5. Reduction in non-revenue water losses

Non-revenue water loss refers to the produced water that is lost during distribution without generating revenue and comprises of components such as real

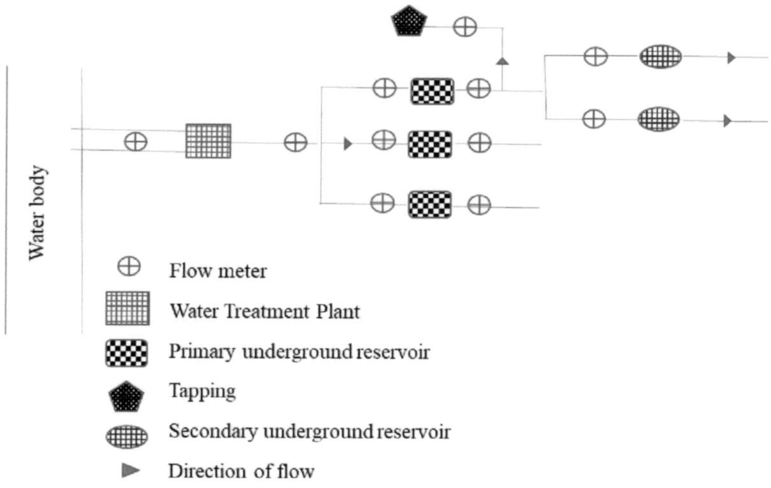

Figure 1.
Installation of flow meters in a water distribution network.

losses, apparent losses and unbilled water consumption [2]. Real losses and apparent losses in a distribution network can be reduced considerably by efficient water auditing. Water flow meters play a critical role in water auditing and give an idea on water losses and usage during each stage. Electromagnetic flow meters can be fitted at different stages of the water supply and a typical scheme of attached flow meters is shown in **Figure 1**. Water from the source water body reaches a Water Treatment Plant (WTPs) and following treatment is diverted to various primary underground reservoirs. Water from the primary underground reservoir can be tapped to various locations and can also be stored in a number of secondary underground reservoirs. Water flow meters installed at different stages can help identify water losses, water consumption patterns and enable better water accounting. The capital city of Delhi for example has Ten WTPs and until 2015 had a total of 350 water flow meters. Sustainable water management practices in the recent years have led to better water accounting and until 2019, a total of 2000 water flow meters have been installed. This has enabled efficient distribution of available water resources and decreased water losses during distribution and water theft.

6. Conclusion

Sustainable management of water resources to provide safe drinking water to all and to protect the natural reserves is one of the major goals of the SDGs put forth by the United Nations. In the current scenario of climate change and water scarcity, it is of utmost importance to strategize urban water management focusing of water use and reuse, conservation of available water resources and sustainable plans to reduce water losses. Sustaining a low water footprint can effectively reduce water scarcity. The developing world is already taking considerable measures to achieve sustainable use of available water resources while many nations in the developing world have started taking initiatives to adopt sustainable water management. Sustainable water management practices can go a long way in protecting the available water resources and preserving the fragile ecosystem for the future generations.

Author details

Prathna Thanjavur Chandrasekaran
Department of Irrigation and Flood Control, Government of National Capital Territory of Delhi, India

*Address all correspondence to: prathna.tc@gmail.com

References

[1] Water Resources Management [Internet]. 2017. Available from: https://www.worldbank.org/en/topic/waterresourcesmanagement [Accessed: 2020-09-09]

[2] Al Washali, T.M., Elkhider, M.E., Sharma, S.K., Kennedy, M.D. A review of non revenue water assessment software tools. 2020. WIREs Water. 7 (2).

Chapter 2

The Availability of Water in Chile: A Regional View from a Geographical Perspective

Javier Lozano Parra, Manuel Pulido Fernández and Jacinto Garrido Velarde

Abstract

Chile is famous for being the longest country in the world from north to south. It means it ranges from polar to desert conditions, water being one of the main limiting factors. In fact, Chile stores a high amount of water (695 mm y^{-1}), but people are not located in the regions where water is more abundant (e.g. in the south). This territorial imbalance is accompanied both by a global context of climate change in which water will be presumably scarcer and by the effects of the current economic activities that are progressively more demanding in water consumption. In this work, we have compared both the current and future availabilities of water for the different regions of Chile in order to provide relevant and useful information on the water balance for land planners. The Metropolitan and Valparaíso regions (Mediterranean climate) along Antofagasta, Atacama, and Tarapacá regions (desert climate) showed the lowest mean values of water availability from 1970 to 2000 (<125 m^3 person y^{-1}). In addition, both the optimistic and pessimistic projections for 2050 forecast a significant increase in the aridity of these two central regions, where the crucial axis between the two most important cities (Santiago and Valparaíso) is located.

Keywords: territorial imbalances, climate change, dynamics, metropolitan areas, sustainability

1. Introduction

Water satisfies several key roles for environmental sustainability and human development: it guarantees the people health and economic development and constitutes the backbone of ecosystems. However, it is irregularly distributed in space and time and frequently is carelessly used to the development of economic activities. According to different economic sectors, just agriculture uses, on average, more than 70% of the water resources in the world to produce food. Thus, the variation of climate patterns will necessarily have an impact on the food supply and its price.

The evidence that mankind faces hydrological and environmental issues is increasingly frequent and visible to the society. This became especially clearer after the last UN Climate Change Conference COP 25, since several countries (such as Spain or France) have declared the climate emergency. Despite this, the results of this meeting have been criticized by several scientific sectors because of their lack of ambition, so that it is valuable to continue making visible the importance of water resources studies and how they will be affected by future climate variations.

In a national context, available water should not be a problem for Chile since it is one of the countries with the highest amounts of water per person [1]. From the year 1970 to 2000, Chile averaged an annual rainfall of 1006 mm (main input) and evapotranspiration of 311 mm as main output, so it has 695 mm of surplus [2]. In terms of storage, it has supposed to be more than 500 km^3 of water susceptible to be used in human activities every year.

Nonetheless, this significant global amount hides a great spatio-temporal variability depending on natural and socio-economic features of each region. A good example of this contrast between regions is the existing difference between the southern regions where water is abundant but people are few and the northern and central ones where population densities are too high in comparison to the availability of water.

There are many situations in which natural conditions are combined with factors, such as economics, to give place to a long water crisis in Chile [3]. For example, if economic activities by sectors are considered, the largest user of water resources in Chile is agriculture, which consumes up to 75% of the water resources at a national scale [4]. This could be justified because it supplies water for an irrigated land of larger than 1 million ha, which is mostly located in the central zone of Chile [5]. Irrigated areas have significantly increased the use of water resources in the last decade. Besides, some products have increased the area for its cultivation.

Among themselves, the crops of avocado (*Persea americana*) have increased 43% in terms of land surface during the last decades, Chile being the second producer country in the world. Avocado is indeed a species that records a water print of 715 l/kg exported, significantly higher than more traditional crops [6]. Most of the Chilean economy is based on natural resources, with a strong export-focused activity [7]. Despite this, in many regions of the country, the uncontrolled exploitation exceeds the actual availability of water resources. This has led to declare numerous regions as depleted in both surface and groundwater [4].

This situation could be even more worrying according to the effects of climate change foreseen by the scientific community. In the particular case of Chile, a reduction between 20 and 40% of total rainfall has been already reported [8]. Added to this, the lack of vegetation management in the upper areas of the watersheds and the elevation of the isotherms have increased water demand and reduced snow deposits, respectively, giving place to a strong decrease of river discharge. An intensive demand for water resources and a decrease in them requires knowing the water availability both in the present and in the future, in order to develop hydrological plans to guarantee economic and environmental sustainability.

Taking into account the abovementioned, this work is focused on the analysis of both the current and future water balance of each one of the administrative regions of Chile since they are the appropriated scale in which decisions on land planning should be made. The reasons of choice this scale work is twofold; on the one hand, Chile is divided into longitudinal regions easily assignable to climatic belts, and, on the other hand, they are the regional governments who are in charge of the regional administrations and consequently in charge of their land and water management. In addition, this research should be also useful to become aware of the upcoming effects of climate change, foreseen by reliable predictive models in every scenario (optimistic and pessimistic), that serve to discuss about the sustainability of many of the current human activities.

2. Material and methods

2.1 Study area

The study was carried out in the whole continental territory of Chile, excluding its islands and the Chilean Antarctic Territory. The size of the study area is 4270 km of distance from north to south and a width (from west to east) ranging from 90 to 445 km involving a great sort of different climate types (cold, temperate, Mediterranean, desert, and high mountain). The limits of the Continental Chile are Peru and Bolivia in the north, the Andes in the east, the Pacific Ocean in the west, and the Drake Passage in the south, Argentina being the country in whom Chile shares more kilometers of border.

This territory is inhabited by around 18 million people with a relatively high standard of living (according to the main macroeconomic indicators) within the context of Latin America. It is divided since 2017 into 16 regions, 56 provinces, and 346 communes. Each region has its own regional government headed by the intendant and the member of the Regional Council (elected every 4 years). These regions keep its original Roman number from I to XII following a gradient from north to south. The Metropolitan Region has the number XIII; and three new regions recently created have continued this tradition (**Figure 1**).

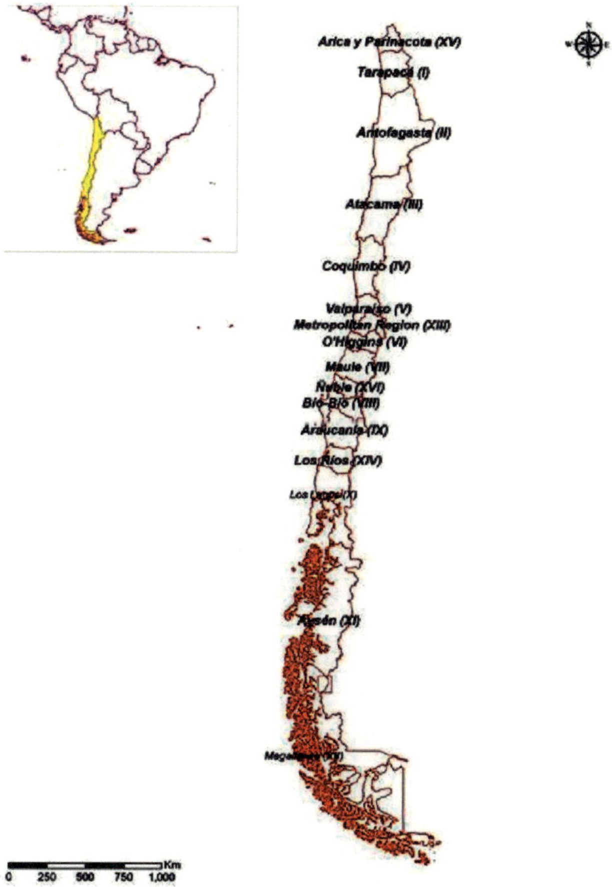

Figure 1.
Geographical distribution of the Chilean regions.

2.2 Water balance

The water balance by region at an annual scale was estimated considering precipitation as the main input and evapotranspiration as the main output of the system, i.e. a positive surplus means precipitation > evapotranspiration and a negative surplus evapotranspiration > precipitation. These data were obtained from Fick and Hijmans [2] for the period 1970–2000 that is considered as the reference period to study changes in climatic variables. Evapotranspiration was determined by using the Turc method [9]. The total land surface of each one of the 16 regions (used to estimate the total amount of water expressed in km^3) and their population (used to quantify the total amount of available water per person) was sourced by Chilean official statistics. Since Ñuble Region (XVI) was officially created in September 2018, its values have been considered within the region of Bio-Bio in which Ñuble was previously included.

2.3 Predictive models

The predictive models utilized in this work were proposed by the Intergovernmental Panel on Climate Change (IPCC) in 2013 [10]. Within those global models used in the fifth stage of the inter-comparison of coupled models, we have chosen the predictions given by the model MIROC5 for 2050 because it has been successfully tested in neighbor countries such as Peru [11]. We have considered two scenarios: (a) RCP 4.5 that assumes an increasing trend in the concentration of greenhouse gases (GHG) until 2040 (the most optimistic) and (b) RCP 8.5 that assumes an increasing trend in GHG concentration for the whole twenty-first century (the most pessimistic).

3. Results

3.1 Regional differences in water availability

Figure 2 shows the mean values (1970–2000) of water surplus per person for each one of the Chilean regions. The lowest values have been recorded in the Metropolitan Region (due to its high population density) and the desert regions of Antofagasta y Tarapacá. Contrariwise, the highest values were observed in Aysén, Magallanes, and Los Lagos. The regions of the north of Chile show relatively low values of annual water comparing inputs by rainfall and losses by evapotranspiration. Of particular interest are the central regions of the country where, on the one hand, the dominant climate is Mediterranean (naturally erratic) and, on the other hand, the pressure for water resources is particularly higher (the Metropolitan region of Santiago is inhabited by more than 7 million people, for instance).

In fact, the highest spatial variability was observed in the central regions of the country where, regardless people and their activities, they show a significant contrast in natural water surplus. These differences start to be visible in the Region of Valparaíso (V) where climatic conditions change to Mediterranean and evapotranspiration is significantly reduced. The five northern regions of Chile (Arica and Parinacota, Tarapacá, Antofagasta, Atacama, and Coquimbo) are scarce in precipitation (<250 mm y^{-1}), and their evapotranspiration is also high (>600 mm y^{-1}), but their needs in water keep in consonance with other regions because their population density is less than 20 people km^{-2} in the best of the cases. These values contrast to the Metropolitan Region (462 people km^{-2}) and Valparaíso Region (111 people km^{-2}).

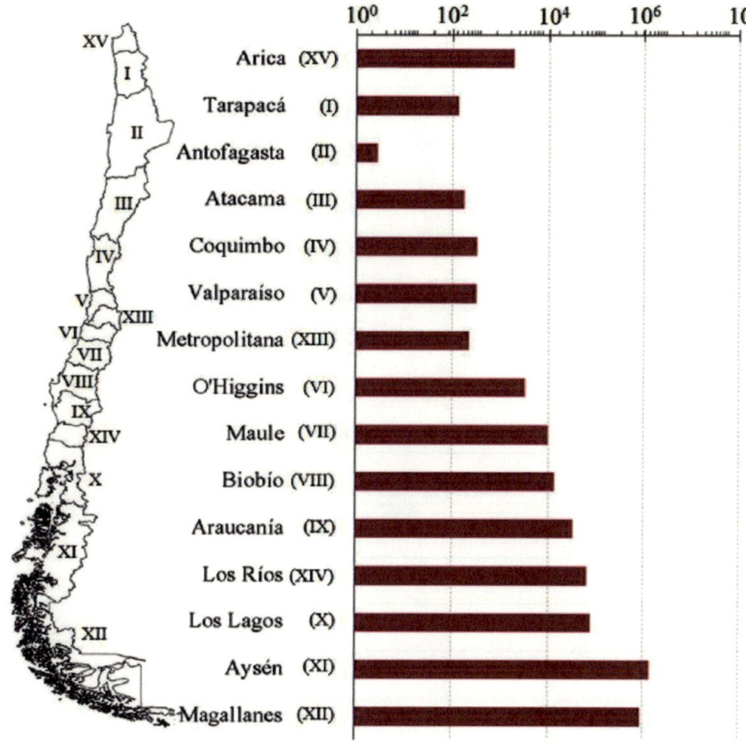

Figure 2.
Regional distribution of the mean annual values of water surplus in Chile (period: 1970–2000).

3.2 Future scenarios

The (optimistic and pessimistic) predictions considered in this research returned similar values for 2050. It means the effects of climate change will be particularly remarkable in the central regions of Chile, characterized by their Mediterranean climate type and their high population densities. Both scenarios foresee the highest decrease in annual precipitation (**Figure 3A**) and increase in evapotranspiration (**Figure 3B**). Regarding water surplus they forecast annual losses of about 1000 mm y^{-1} in many regions of the country (**Figure 3C**). These losses are particularly worrying in the centre where most of Chilean people are now living.

The existing differences between both scenarios (optimistic vs. pessimistic) are not significantly different between them. For instance, the most optimistic foresees a reduction of 936 mm y^{-1} in precipitation and the pessimistic one of 1052 mm y^{-1}. Regarding water surplus, both scenarios forecast losses around 1000 mm y^{-1} (optimistic, 968 mm y^{-1}, vs. pessimistic, 1094 mm y^{-1}) regardless of the number of people and the water consumption of their agricultural activities. In the south of the country, the model returns predictions of an increase in precipitation much higher than the foreseen increase in evapotranspiration. It means much more available water will be in the south, and perhaps it can suppose migrations from the north and central of the country to these regions. Regarding the desert regions of the north of Chile, the future situation will be presumably similar than the current one. So, no significant effects on local population are expected.

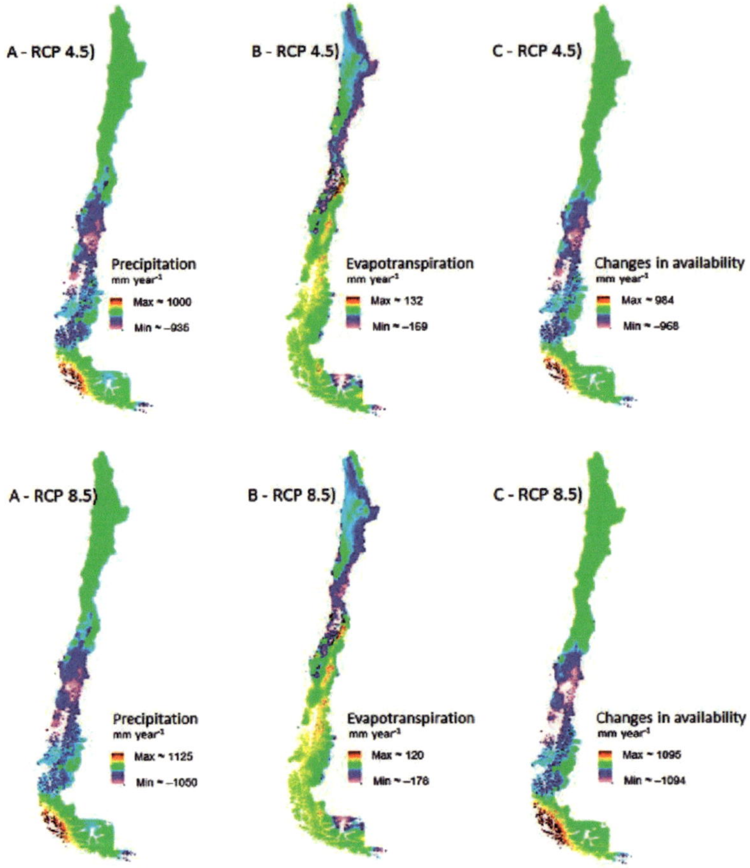

Figure 3.
Differences between the averaged values of the period 1970–2000 and those foreseen by the predictive models for 2050.

4. Discussion

Although the effects of climate change on the water availability of Chile are well-known, or well-guessed, by the majority of people, their needs and demands in water are continuously increasing. For instance, the official organization in charge of water management (DGA, Dirección General de Aguas in Spanish) is still granting new rights for water exploitation in the regions of the north of Chile in spite of the thresholds of natural recharge of the aquifers having been already exceeded [5]. In addition, in this work we are not addressing further effects of this water scarcity such as salinization, really negative in semiarid countries such as Algeria [12].

In the north of Chile, the use of water is related to mining activities. The price of copper has significantly increased and it has supposed catalysis in mining activities, i.e. a much higher consumption of water by these companies [4]. Other sectors such as the energetic (e.g. hydroelectric) are also intensifying their pressure on water resources since they are now living an expansionist time [3], perhaps provoked by many decades of criticism against the use of fossil energies such as petrol and coal or the pressure exerted by conservationists that consider wind turbines to be dangerous for birds [13].

Regarding the agricultural sector, water is progressively being more demanded because of the growth of population in Chile and the increase in the amounts of products exported to the global market [14]. It is provoking a higher scarcity and more competition for water in the whole country. In fact, Larraín and Poo [3] have reported a worrying increase in the number of conflicts related to (spatially unequal) water management. Most of these conflicts are happening in the central and north of Chile where water has become a valuable good due to erratic climatic conditions (Mediterranean and semiarid climates) and increasing pressure of the economic sectors [4].

Water management will be a challenging question for any regional and national government of Chile. The reserves of water in the aquifers and the estimation of the accurate consumption of water by people, mining, and different commercial crops such as avocado will decide the sustainability of this valuable resource. Specifically, in the agriculture, the recent crops (thought for exportation) are important consumers of water. In fact, most of them grow naturally in tropical conditions, and they are produced in Chile by using irrigation water [15].

The branch of knowledge, with an outstanding global success, that is dealing this problematic is being the ecohydrology, recently popularized and key as scientific support for making decisions on water and land management properly by the political power [16]. So, we consider countries at risk of having serious problems of water availability; Chile, Spain, and Israel, for instance, should pay more attention and promote more specific research programmes for ecohydrologists. Studies on land capability of some crops and on the sustainability of traditional systems such as the *espinal* we think are more necessary than ever.

Another challenging task both for scientists and decision-makers is the proper scale of work. Here we have emphasized about the existing differences between the 16 administrative regions of Chile, but it is totally necessary to know the mechanisms that rule in a leaf, plant, farm, and catchment, among other work scales. In addition, the rights for water exploitation granted by the DGA should be very strict in terms of the total amount of water per year that each company must utilize. These rights should vary in time according to the rainfall variability or presumably increasing in evapotranspiration, i.e. companies and/or public agencies should have tools for monitoring water reserves constantly.

Another aspect that can be also discussed by the scientists and decision-makers, although it is not the *leitmotiv* of this work, is the usefulness of using long-term datasets. It means an extraordinary effort in monitoring many parameters linked to water and vegetation is still needed since datasets of more than 10 years are still infrequent [17]. Some published works based on long-term datasets [18, 19] have contributed to understand the interrelationships between ecohydrological processes and the time of adaption with respect to the arrival of environmental changes.

In this study, we have shown the variation in water surplus for 2050, but nothing has been mentioned about the synergetic interaction between water and vegetation. The presence of vegetation is a cause and consequence of the presence of water and vice versa [15]. For instance, the spatial vegetation patterns exert a strong control on the spatio-temporal variability of soil moisture and, consequently, on water infiltration or redistribution of water through soil macropores [20]. In addition, it is soil moisture content who determines species phenology and biomass production [21].

The works published on this topic in the last decades have served to understand that the existing relationships abovementioned between water and vegetation parameters are nonlinear. According to Scheffer et al. [22], these relationships are based on thresholds, i.e. some properties keep on a stable status until they cross a certain threshold. It means that ecosystems of Central Chile that traditionally have been Mediterranean (dry subhumid) can change to semiarid and keep steady for decades and centuries. Nonetheless, the concept of stable status has just arrived to the studies of ecohydrology. So, further research is still necessary to draw definitive conclusions.

Every ecosystem is controlled by the availability of water that it is naturally provided, but it can be worn up by human activities. In the ecosystems in which water is naturally abundant, the control of ecohydrological processes is ruled by the level of the water table [15]. Meanwhile in Mediterranean and semiarid ecosystems, ecohydrological processes are adapted to different moment or pulses ruled by water surplus or deficit in the non-saturated soil zone [23]. To investigate these mechanisms and processes to better understand the sensitivity of each ecosystem face environmental disturbs is the main challenge of the ecohydrology.

5. Conclusion

The availability of water in Chile is of crucial interest for the near future since it is one of the countries that will be more affected by climate change. The regions that show a higher risk are those located in the central part of the country that are ruled by dry subhumid conditions (Mediterranean climate type) and inhabited by almost half of the total population. This forecast should serve the decision-makers to prepare a new scenario in which high consuming crops in water such as avocado (*Persea americana*) and mining activities perhaps cannot be exploited due to lack of water. Another controversial question that should be treated in the future is the role in water management of the DGA (*Dirección General de Aguas*) since it is a national agency and the regional differences detected can be a good indication of the necessity of making decisions at a regional level, i.e. by the regional governments of each one of the current 16 regions of the country. Finally, more efforts in data collection are still needed both by part of the scientists and by governments that grant research programmes, funds, and scholarships.

Acknowledgements

This study has been possible by funding provided by the Fondo Nacional de Desarrollo Científico y Tecnoóógico [FONDECYT 11161097] and by the Programa de Cooperaóión Internacional para la Formóción de Redes Internacionales de Investióación [REDI170640], both granted by the Government of Chile.

Author details

Javier Lozano Parra[1]*, Manuel Pulido Fernández[2] and Jacinto Garrido Velarde[2]

1 Pontifical Catholic University of Chile, Santiago, Chile

2 Research Institute for Sustainable Territorial Development, Cáceres, Spain

*Address all correspondence to: jlozano@uc.cl

References

[1] Lajaunie M, Scheierling S, Zuleta J, Chinarro L, Vazquez V. Chile: Diagnóstico de la gestión de los recursos hídricos. Washington, DC, USA: World Bank; 2011

[2] Fick SE, Hijmans RJ. WorldClim 2: New 1-km spatial resolution climate surfaces for global land areas. International Journal of Climatology. 2017;**37**:4302-4315

[3] Larraín S, Poo P. Conflictos por el agua en Chile: Entre los derechos humanos y las reglas del mercado. Santiago de Chile: Heinrich Böll Foundation; 2010. pp. 362

[4] Rivera D, Godoy-Faúndez A, Lillo M, Alvez A, Delgado V, Gonzalo-Martin C, et al. Legal disputes as a proxy for regional conflicts over water rights in Chile. Journal of Hydrology. 2016;**535**:36-45

[5] Hearne R, Donoso G. Water markets in Chile: Are they meeting needs? In: Water Markets for the 21st Century. Springer; 2014. pp. 103-126

[6] Juárez Laguna YF. Impacto del cambio climático en la disponibilidad de agua para el cultivo de palta. Lima: Universidad Científica del Sur; 2019

[7] Valdés-Pineda R, Pizarro R, García-Chevesich P, Valdés J, Olivares C, Vera M, et al. Water governance in Chile: Availability, management and climate change. Journal of Hydrology. 2014;**519**:2538-2567

[8] Gironés Mompó FJ. Evaluación de las predicciones climáticas de modelos globales en Chile. Valencia: Universitat Politècnica de València; 2019

[9] Turc L. Estimation of irrigation water requirements, potential evapotranspiration: A simple climatic formula evolved up to date. Annales Agronomiques. 1961;**12**:13-49

[10] Intergovernmental Panel on Climate Change. Climate Change 2013: The Physical Science Basis: Working Group I Contribution to the Fifth Assessment Report of the Intergovernmental Panel on Climate Change. New York: Cambridge University Press; 2014

[11] Moya Álvarez A, Ortega León J, Jurado PX. Evaluación del Modelo Climático Global MIROC5 y estimaciones de temperatura y precipitaciones para las zonas sur y norte del Perú. Apuntes de Ciencia & Sociedad. 2015;**5**:188-195

[12] Abdennour MA, Douaoui A, Bradai A, Bennacer A, Pulido FM. Application of kriging techniques for assessing the salinity of irrigated soils: The case of El Ghrous perimeter, Biskra, Algeria. Spanish Journal of Soil Science. 2019;**9**:105-124

[13] Sovacool BK. Contextualizing avian mortality: A preliminary appraisal of bird and bat fatalities from wind, fossil-fuel, and nuclear electricity. Energy Policy. 2009;**37**:2241-2248

[14] Rodell M, Famiglietti J, Wiese D, Reager J, Beaudoing H, Landerer FW, et al. Emerging trends in global freshwater availability. Nature. 2018;**557**:651-659

[15] Rodriguez-Iturbe I, D'Odorico P, Laio F, Ridolfi L, Tamea S. Challenges in humid land ecohydrology: Interactions of water table and unsaturated zone with climate, soil, and vegetation. Water Resources Research. 2007;**43**:1-5

[16] Eagleson PS. Ecohydrology: Darwinian Expression of Vegetation Form and Function. Cambridge University Press; 2005

[17] Lewis D, Singer M, Dahlgren R, Tate K. Hydrology in a California oak woodland watershed: A 17-year study. Journal of Hydrology. 2000;**240**:106-117

[18] Asbjornsen H, Goldsmith GR, Alvarado-Barrientos MS, Rebel K, Van Osch FP, Rietkerk M, et al. Ecohydrological advances and applications in plant–water relations research: A review. Journal of Plant Ecology. 2011;**4**:3-22

[19] Seneviratne SI, Corti T, Davin EL, Hirschi M, Jaeger EB, Lehner I, et al. Investigating soil moisture–climate interactions in a changing climate: A review. Earth-Science Reviews. 2010;**99**:125-161

[20] Lozano-Parra J, Schaik NLMB, Schnabel S, Gómez-Gutiérrez Á. Soil moisture dynamics at high temporal resolution in a semiarid Mediterranean watershed with scattered tree cover. Hydrological Processes. 2016;**30**:1155-1170

[21] Lozano-Parra J, Schnabel S, Ceballos-Barbancho A. The role of vegetation covers on soil wetting processes at rainfall event scale in scattered tree woodland of Mediterranean climate. Journal of Hydrology. 2015;**529**:951-961

[22] Scheffer M, Carpenter S, Foley JA, Folke C, Walker B. Catastrophic shifts in ecosystems. Nature. 2001;**413**:591-596

[23] Lozano-Parra J, Maneta MP, Schnabel S. Climate and topographic controls on simulated pasture production in a semiarid Mediterranean watershed with scattered tree cover. Hydrology and Earth System Sciences. 2014;**18**:1439-1456

Chapter 3

Smart Water—How to Master the Future Challenges of Water Management

Günter Müller-Czygan

Abstract

Innovative digital developments from industry like autonomous machine controls based on intelligent data acquisition, collection and evaluation, promises better adapting municipal infrastructure systems to changing conditions. When the technology initiative KOMMUNAL 4.0 was developed as an idea in 2015, digitalization was not a central topic in water management. As Industry 4.0 was present everywhere in the media, the idea of transferring suitable parts of the basic idea of Industry 4.0 to municipal water management was born. In particular, it was necessary to implement consistent IT and IoT communication at all levels of water management tasks. The aim was not only to create a uniform structure for networking a wide variety of applications, but also to round off KOMMUNAL 4.0's complete range of services with IoT for existing and newly developed products and solutions. Regardless of whether it concerns measurement and data technology applications, smart machines, SCADA or asset management systems, all application solutions contain a standardized network core that guarantees standard data communication and also complying with safety and cybersecurity requirements.

Keywords: digitization, smart water, municipal 4.0

1. Introduction

In Germany, approx. 6–7 billion EURO is invested every year in the renovation or new construction of buildings and plant technology in the municipal water and sanitation sector [1]. The German water and wastewater infrastructure has developed socially and spatially balanced in the past and has grown over many decades and guarantees today a comprehensive disposal with high drainage safety combined with an extremely long technical and economic service life. In opposite it results in a lack of operational flexibility for sewer network and sewage plant operators, e.g. in the event of extreme weather events as an effect of climate change, changed consumer behavior or the consequences of demographic change. Experts and decision-makers are therefore looking for ways to adapt the dimensioning and calculation of future investments more closely to real usage requirements and to dispense with previous inaccurate estimates. At the same time, the existing systems must be operated more flexibly and thus more efficiently, even under described changed conditions.

Innovative digital developments from industry like autonomous machine controls based on intelligent data acquisition, collection and evaluation, promises better adapting municipal infrastructure systems to changing conditions. When the technology initiative KOMMUNAL 4.0 was developed as an idea in 2015, digitalization was not a central topic in German water management. As Industry 4.0 was present everywhere at these time the idea of transferring suitable parts of the basic idea of Industry 4.0 to municipal water management was born. In particular, it was necessary to implement consistent IT and IoT communication at all levels of water management tasks (**Figure 1**). The aim was not only to create a uniform structure for networking a wide variety of applications, but also to round off KOMMUNAL 4.0's complete range of services with IoT for existing and newly developed products and solutions. Regardless of whether it concerns measurement and data technology applications, smart machines, SCADA or asset management systems, all application

Figure 1.
Consistent IT and IoT communication of digital products and systems [2] (translation: IT-Sicherheit, Asset Management und Digitalisierung, Betriebsführung, Recht = IT security, asset management and digitization, operations management, law; Fernwirkung, Fernüberwachung = remote control, remote view; Automatisierung, software = automation, software; mess-und Datentechnik = measurement and data technology).

solutions contain a standardized core that guarantees standard data communication and also complying with safety and cybersecurity requirements.

Another important requirement was (and is) that all applications work as individual and independent solution. This enables the user to go down the path to digitization in individual steps, which are, however, coordinated with each other right from the start. The purchase of a complete system at the beginning of a digital process is not absolutely necessary. The user can start where there is currently the most urgent need at daily work without losing the network compatibility of individual elements that have to be adding later. One of the most important tasks in municipal water management, for example, is an effective and efficient management of the entire infrastructure. Data plays an increasingly important role at this topic. Only where data from different sources can be usefully related to each other real added value can be created. Various IT systems such as GIS, process control (SCADA) or asset and maintenance management systems are used for this purpose in water management. Systems are desirable which, like MS Office, function in both ways as individual solutions and offer high benefits by networking with each other. And just as every printer today communicates perfectly with MS Office smart products, measurement and data technology applications as well as Smart Machines should be integrated in a plug & play manner. Some of the products and solutions belonging to KOMMUNAL 4.0 already offer these requirements already today.

2. Industry 4.0—a model for sustainable water management?

The digitization offensive of the industry, known since 2013 in Germany as Industry 4.0 and initiated by the Federal Government, is intended to turn simple machines using the Internet into so-called Smart Machines. These are self-regulating production units (they are also called CPS = Cyber-Physical Systems) which leads to significant cost savings. For example, they are fed with orders directly from commercial databases, receive their technical instructions directly from CAD/EPLAN tools of development engineers, order necessary materials independently from suppliers, coordinate their interdependencies and report the completion of the manufactured products to logistics for dispatch. The entire industrial value chain is recorded in data form, analyzed and controlled or optimized by automatic processes. Can this approach be transferred to the level of water infrastructures as a model?

In water management, the possible applications of intelligent and smart solutions are being intensively discussed and are already being used (see e.g. at [3–6]). Modern automation technology for water management already has elements in its core that need not fear comparison with Industry 4.0 solutions. Real-time-based control or monitoring solutions are just as much in use as numerous intelligent sensor technologies. They form an important basic framework for future digital strategies. In order to obtain innovative and thus sustainable digitization solutions, such automation and IT systems must be extended by suitable analysis and evaluation tools (Big and Smart Data). Only this enables an intelligent networking of several objects with each other.

For the municipal user, the question now arises with whom he can start digitization. There are many specialist providers for individual application solutions, but how will be done a well integration into a future platform solution? Whoever is faced with the procurement of new IT systems, e.g. in the GIS/PLS-SCADA/ERP/BFS areas, that is not an easy task to master. If there is a high degree of network compatibility due to a close technological relationship between the individual solutions (see example MS Office), this facilitates the start incl. a step-by-step development of a complete digitization.

In the fields of municipal water management that are eligible for digitization, there are already a large number of established providers whose solutions in principle include these useful and expected functions. With regard to the basic functions, normal companies are moving forward in small steps. The differences in the functions can be recognized and evaluated more precisely after intensive use. If providers have several applications, there is concern on the user side not to obtain the supposedly best software for every task. If, however, a provider understands the current requirements of networking and has its own development of its product lines, this also has significant, clear advantages for the user compared to the sole availability of a special function. These range from interoperability up to the elimination of un-useful complex parameterizations. The importance of standardized user interfaces and dialogs, administration, data formats, menu navigation, etc., can be seen again when considering the development of the office world. For the KOMMUNAL 4.0 product lines, the overall system selection is based on the aspects of secure investment through technological sustainability as well as networking and usability. In this case, the technology is concerned with the methods, technologies and resources used in product development itself. They are essentially responsibilities for what is working today, in the future and what does not fit. At this point the user must inform himself accordingly early enough in order not to wait too long or also in vain for the necessary adaptations of his (special) provider in the future.

In particular, the currently pending change through digitalization is a good way to orient himself comprehensively and to make new decisions if necessary. If the user succeeds in procuring systems from one platform and possibly from one provider, the networkability and operability of the overall solution will be simplified. The solution should also offer the possibility to integrate already existing software applications. The conversion and renewal of software and its entire technological basis also requires a lot of orientation, strength, competence and investment at the providers side. A changeover from classic client-server software to web systems, for example, also requires new thinking in development. As a result of the high challenges, only small steps or adaptations of the outer shell are often successful but no innovative progress or better results have to be achieved. Fitting usability and an intuitive using of a software can only be determined after several days of practical work. On the other hand users expect more today. They want deep horizontal and vertical networking of the systems, e.g. with asset/maintenance and SCADA systems. This has taken on a new and higher significance and it is the most discussed topic of interfaces or integration capability or networkability of the systems. Synchronization functions, uniform grammar, file formats, reliability and warranty are more and more in focus. Regarding these requirements the KOMMUNAL 4.0 world is already one step ahead and offers compatible web systems for GIS/PLS-SCADA/ERP/BFS tasks as well as integrated measurement and data technology applications and smart machines. The company HST (consortium leader of KOMMUNAL 4.0) for example has converted all its systems to platforms and web technology in recent years and comprehensively networked its systems. The widely used asset/maintenance management system KANiO and the process control system SCADA.web are today highly communicative networkable solutions with open standard interfaces as well as integration and synchronization functions also for third-party systems. Reliability and optimal operability have priority.

An IT-supported ISMS (Information Security Management System) is already available (KANiO-ISMS) for secure compliance with the requirements of the IT German Security Act. It represents an important building block for the individual steps on the way to a legally compliant IT security structure and is available as a separate application and also as a component of the KANiO system. The use of the tool avoids uncoordinated individual measures that do not guarantee sufficiently secure IT operation. In addition, the tool ensures that the company's own efforts to ensure secure IT operation can be proven to customers or legislators. Earlier measures can thus also be better aligned with the current security standard. In addition, almost all process engineering machines of HST are gradually being equipped with sensors and actuators (so-called IntelliSystems) so that SCADA and asset/maintenance management system can be directly networked and collaborated. This means that there are already solutions for integrated IT and IoT communication across all application levels, as shown in **Figure 1**.

3. Why does digitization differ from earlier technology developments?

Classical engineering-based research does not fully reflect the comprehensive developments of digitization. With digitization, something very big has happening with increasing speed. It has now also reached the water industry. An analysis of publications, studies and research carried out in the context of KOMMUNAL 4.0 in the field of industry 4.0 has shown that, in addition to technology patterns, other subject areas are important which will also have a serious impact on the water management. This applies, for example, to data sovereignty, data law and public procurement law.

Thanks to the accompanying research of the federal technology program "Smart Service World", in which the KOMMUNAL 4.0 project is embedded, and the associated networking with other Industry 4.0 research networks such as the federal technology program "Smart Data", the project partners in KOMMUNAL 4.0 are able to access the current state of knowledge and expected developments not only at the field of IT security and legal issues. These coming topics, some of which have not even been discussed in the water industry until today, are already the subject of KOMMUNAL 4.0.

Previous technology flows primarily arose from development ideas that were examined in individual objects or tasks. An example of this is the extended elimination of nitrogen in sewage treatment plants. When this aspect was investigated and applied broadly after being anchored in legislation, the effects were limited to the respective sewage treatment plant or the responsible organization. The other departments of a municipality, city or association were not affected. The same can be noticed for example for the so called fourth sewage treatment stage. From this point of view, research institutes and plant operators were able to investigate into such issues independently to get an isolated developed solution. Consequences of a lack of communication between research institutes were not significant due to low need of interdisciplinary interfaces. Although the existing scientific-technical exchange among experts is maintained and also inspires research initiatives, a mutual agreement on the research and development contents was not absolutely necessary for the previous form of technology development.

In opposite to aforementioned situation the appearance of digitization must be judged differently. Since as a cross-sectional technology it has an almost unlimited influence on all technological and organizational environments. At the horizontal level individual objects such as rain basins, flood reservoirs, water treatment plants or sewage treatment plants have to be connected more and more with the entire infrastructure systems across city and municipal boundaries. At the organizational level (vertical level) different departments and organizations have to be linked to each other. In addition, regional and supra-regional administrative and authority units should also be integrated into these networking efforts.

"It also applies to water management that everything that can be digitized has to be digitized". This comment made by Martin Weyand, BDEW Managing Director Water/Wastewater [7] confirms the cognitions of KOMMUNAL 4.0. Previous technological developments in water management have had only a limited impact on organizations and working methods, but in contrast to this, digitalization is expected to bring about massive changes in the everyday working lives of individuals as it unfolds its full potential. Already from individual elements as part of growing complex systems more far-reaching effects are to be expected. This leads to the conclusion that an examination of new digital solutions must be more comprehensive and must go beyond the previous horizon of knowledge and understanding. This is where the holistic approach of KOMMUNAL 4.0 comes in, in which all relevant individual modules and their interaction with each other were taken into account from the very beginning. This approach allows a better assessment of how to deal with digitization, even if it increases the amount of work at the beginning. As a result, it is easier to assess the major consequences of individual solutions and thus enables the foresighted engineering of networked systems. Based on this, current and future products and solutions will be manufactured.

4. The KOMMUNAL 4.0 project—a German beacon of digitization

4.1 The funded project KOMMUNAL 4.0

The cooperation project KOMMUNAL 4.0, which is funded by the German Federal Ministry of Economic Affairs and Energy, will devote itself in a special way to the challenges described above. Current and expected future developments in the field of Industry 4.0 were examined for applicability in municipal water management. KOMMUNAL 4.0 was selected as one of the 16 winners in a pure industrial competition from 130 applicants [8]. The intended developments for digitization lead to higher efficiency, safety and control in the operation of water management plants and systems and can serve as a model for other infrastructure sectors. The current low adaptability of municipal infrastructures to changing conditions such as heavy rainfall or demographic change can be significantly increased with the help of the IT and organizational solutions from KOMMUNAL 4.0.

The project consortium (see www.kommunal4null.de) under the cooperation management of HST Systemtechnik GmbH & Co. KG strives for the following essential goals:

- Standardization of data acquisition and transmission from heterogeneous CPS (cyberphysical systems).

- Development of a web-based data platform for collection, structuring and conversion of different data/data formats.

- Development of flexible platform architecture for optional use as intranet or internet application.

- Development of application tools in the areas of design/engineering, benchmarking, object/network monitoring, data fusion, procurement, end-to-end process chain and operational optimization.

- Development of required IT security concepts.

- Development of digital business models such as machine sharing.
- Analysis of legal aspects of cloud computing.

The developments focus on modular and step-oriented solutions. It starts with individual intelligent aggregates, so-called smart machines, and lead to the linking of several objects with each other up to a completely networked infrastructure system. Essential objects of the project are so-called pilot projects. Here, the developed application tools were installed at selected municipalities or operators in a real infrastructure environment incl. comprehensively testing.

4.2 Association KOMMUNAL 4.0 e.V

In order to maintain the previous ideas and the already established network of experts in the KOMMUNAL 4.0 funding project after the end of the project (31.12.2019), the Federal Association KOMMUNAL 4.0 (www.kommunal4null-ev.de) was founded in 2017. It supports the previous basic and competence transfer of the funding project and will work as a hub for the digitization of municipal infrastructures. It also takes care of central tasks such as public relations for digitization, training and further education, standardization and networking. The association sees itself as a central point of contact for planning and implementing the first steps toward the digitization of municipal infrastructures or for carrying out advanced technological expansions. Even though the current focus is on the municipal water sector due to the proximity to the funding project, all other relevant sectors of municipal infrastructure are to be added in the future.

A central importance for the water sector is the establishment of a KOMMUNAL 4.0 academy. So far, the sector is not be able to offer any application-related further training courses. The Federal Association KOMMUNAL 4.0 will offer a corresponding service which covers the topics IT security, IT systems, operational management, process control engineering as well as measurement and automation engineering. But there will also be application-related offerings, e.g. how digitization can look specifically in water supply or wastewater disposal or in special structures such as pumps, rainwater basins, sewage treatment plants, etc. In addition, there will be special seminars for mayors, heads of offices and planners so that these industry participants can set their very special requirements in relation to the challenges of digitization.

4.3 From smart machines to smart infrastructures

Embedded systems have been around for a long time at the water management. The state of the art is that mechanical aggregates are connected to automation technology, which takes over monitoring, control and regulation functions based on various information (mostly from measurement sensors). Automation technology is also used for data acquisition and transmission to higher-level units such as SCADA systems. They form an important part of a complete networking solution (see **Figure 1**).

The stored specifications of a smart machine follow clear assignments and rules, especially for control. Changes to the specifications are made by the operator via set points or directly at the PLC level by a programmer. Data is linked locally by cable. And how do smart machines emerge from this? Thanks to the availability of rapidly increasing web-based application options, the monitoring, control and regulation of actuators no longer needs to be carried out in isolation with locally recorded data and locally used automation technology. For example, additional information such as current precipitation data or status information from a piping system can be transferred from a central database to the local controller via a wireless Internet connection or data line. Based on corresponding algorithms, the controller permanently analyzes the functional environment (in real time) and independently adapts the control specifications (set points) to changing environmental conditions. This is illustrated by the example of a pumping station.

Pumps are designed for an optimal but static operating point based at only one expected operating situation. However, fluctuating water volumes and losses due to unfavorable piping or other operating conditions cause pumps to run outside their selected characteristic curve. This is also due to the fact that, unaware of the actual delivery peaks, corresponding safety surcharges/reserves are provided while dimensioning the pumps. This results in higher energy consumption and less efficiency of the overall system and thus also reduces the service life of the units. Innovative pump controllers (software solutions such as so-called IntelliPump system) permanently evaluate the entire operating situation and, by using frequency control, enable operating sequences that permit several optimum operating points depending on the requirements. This permanently guarantees the intended pumping safety and thus reduces wear and energy consumption of the pump. Another advantage is the continuous monitoring of system operation. This enables faults to be detected more quickly and a better condition assessment of the machine, thus increasing overall operational safety. The formerly simple pump becomes a smart pump system.

In the near future, smart machines will become standard equipment in water management, also as a result of the KOMMUNAL 4.0 project. The connection to web-based data portals, such as the precipitation portal NiRA.web, increases the adaptability and efficiency of individual machines and the system in complete. The virtual connection of the machines with the Internet allows access to all operating data from a central location. Selected operation-relevant data supports local machine control, link systems/objects with each other and ensure efficient operation throughout the entire infrastructure system. An example of this is a sewer network with various rainwater basins, pumping stations and a central connected sewage treatment plant.

The interconnection of the objects, as shown in **Figure 1**, permits an optimal congestion, flushing and operating regime of the entire infrastructure network. A central data evaluation of all structures decides about the right time to empty storm water tank, e.g. to keep sufficient storage capacities free for a next heavy rainfall or flood event, or to make optimum use of capacities or to control the relief events from storm water tanks in the sense of optimum water protection. The more quantitative and qualitative data are available for each structure/object, the better and more efficiently each individual machine, each object and also the entire infrastructure system can be operated. Similar applications, e.g. the intelligent basin cleaning system IntelliGrid, the self-regulating occupancy control system IntelliScreen for increasing the material retention in horizontal bar screens or the EMA flow rate recording system at rainwater overflows, are increasingly being used in water management. In the course of the KOMMUNAL 4.0 project, the prerequisites are now being created for networking individual applications across buildings in order to create a genuine, smart infrastructure.

5. Where to start? Start digitizing correctly!

If, for example, current new installations of technical equipment are due, this can be the ideal start of digitization on the basis of individual measures. At this stage, it should be examined whether it makes sense to design the new technical equipment as a so-called smart machine or as a smart system. If digitization is started with a smart local solution, it must be ensured that this solution is also future-compatible with larger networking solutions, such as the KOMMUNAL 4.0 platform. A municipality benefits from this kind of digitization very early, for example by installing a smart machine. This is a comparatively simple way of approaching the complexity of digitization.

Smart machines and solutions based on the Intelli principle work autonomously with the full advantages of digitization and can therefore be easily integrated into a higher-level networking system at a later point in time, even if an overall digital strategy for the municipality has not yet been defined. **Figure 1** shows the systematics of networked products that are already prepared for a platform connection and cover almost the entire range of applications on a horizontal and vertical level. The same applies to upcoming new acquisitions of software solutions in the areas of asset/maintenance management systems and SCADA. The compatibility to the (smart) machine world has to be checked. The necessary knowledge can be acquired by the KOMMUNAL 4.0 experts.

The example of the selection of a computer system on a relief threshold of a sewer system will illustrate how smart systems as described can be applied. Increasingly, screening systems are being used on discharge thresholds to reduce the amount of dirt discharged into water bodies during discharge events. Conventional systems automatically clean the screen bars at fixed intervals. The focus here is on ensuring the hydraulic capacity, regardless of whether the current operating condition requires this or not. Smart rakes equipped, for example, with the IntelliScreen system (see Smart Machines IoT level in **Figure 1**) use networked information from local machine, operating data, webcam data and precipitation data from data portals (see Measurement and Data Technology level in **Figure 1**) to achieve greater operational safety and water protection.

While overflow screens have been cleaned by continuous comb and/or clearing devices up to now, screens equipped with Intelli systems have the advantage of recognizing their current and prognostic screenings. In addition, speed-controlled drives enable variable combing and clearing speeds and extended power reserves. Networking and the inclusion of precipitation data enables an even more accurate prognosis of the operating process and the combing and clearing requirements. On the basis of this expanded and improved information situation, the filter effect of the screenings is now used more intensively and for longer in terms of water protection on the one hand. On the other hand, in the case of heavy rainfall and overflow requirements, the spatial performance and thus the relief safety is increased. The machine works locally by integrating digital precipitation data from a web portal. In further steps, the machines are connected to a process control system (see level Telecontrol or remote monitoring technology in **Figure 1**) or integrated into an asset/maintenance management system for the organization of the necessary maintenance and repair work, in which the documentation requirements of the IT Security Act are also fulfilled by using an ISMS system (see IT Security, Asset Management and Digitization, Operations Management, Law in **Figure 1**).

The integration of the various system modules as shown in **Figure 2** into a data and service platform (e.g. KOMMUNAL 4.0) optimize the technical side of digitization. All data streams flow together at this platform and can be processed for further analyses and purposes such as Big and Smart Data or for operational support with a user-specified dashboard (see **Figure 3**). The system in **Figure 1** can also be used in the form of a process template to derive the necessary organizational measures from the technical elements.

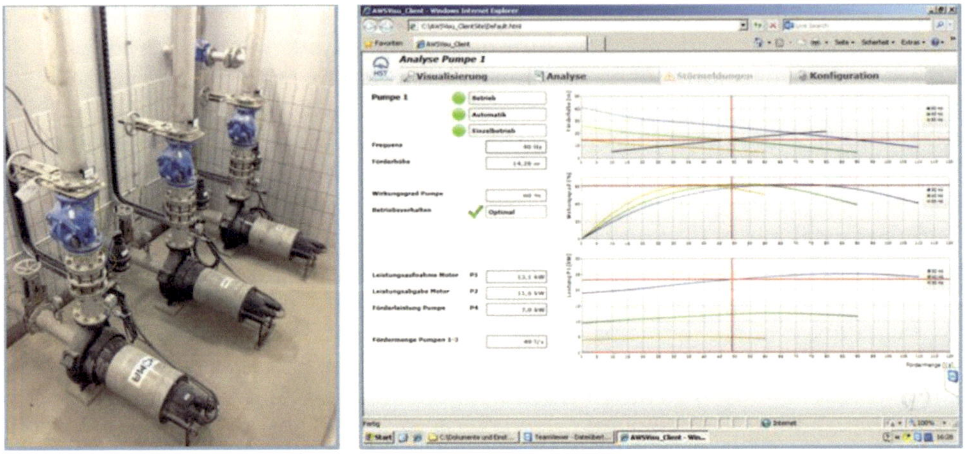

Figure 2.
Increase pump efficiency with IntellPump software [4] (screenshot shows real pump characteristic curve and its adaptation by software to ideal curve).

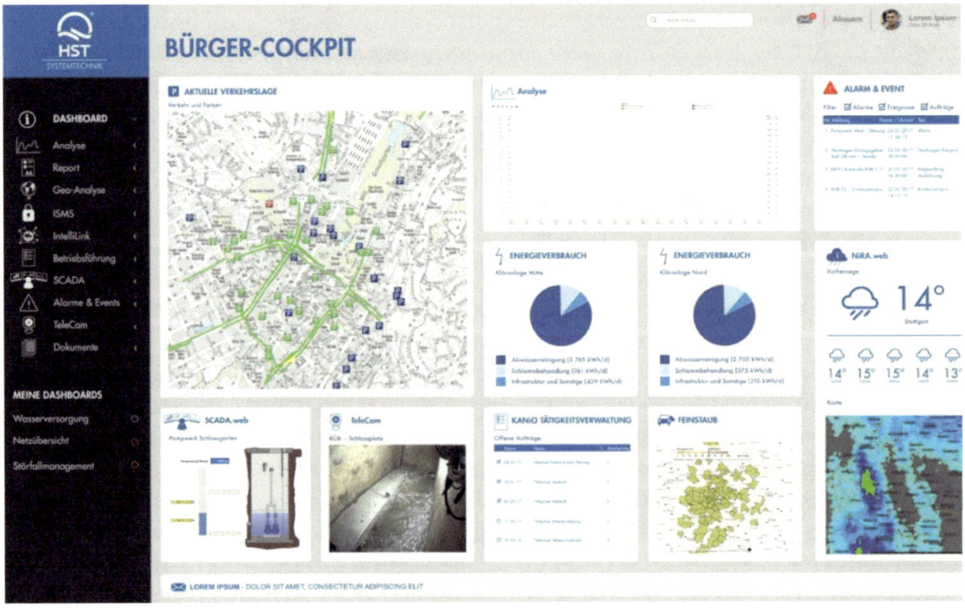

Figure 3.
Dashboard KOMMUNAL 4.0 [9] (screenshot shows example for a KOMMUNAL 4.0—cockpit = cockpit of a smart city. It shows different data monitoring systems of water facilities that includes energy consumptions, water level, traffic, dust, alarm events incl. local weather data).

5.1 Start with "anyway" projects

Even it is often propagated that the development of a comprehensive digital strategy is needed to start digitization, it is often better to start digitalization at a concrete and manageable practical case. Also at the beginning of KOMMUNAL 4.0, the planned application ideas were very strongly described from the perspective of an abstract digitization vision. Addressed municipal users (rightly) hardly understood these ideas and could not transfer them to their own application needs. More and more the communication of the project goals and the first results were changed to take the needs of the municipalities in clear focus. With this strategy suitable digitalization ideas could be discussed and subsequently projected. The most important result to achieve an ideal start was using a current and manageable investment project as an introduction to digitalization [10]. For this purpose, the project partners carried out an analysis of a possible "Anyway" project (investment project, which has already been determined for implementation) and examined how a KOMMUNAL 4.0 solution would serve the respective project objective. In many cases, individual measures have to be filtered out from these "Anyway" projects, in which digitization could be tested to a manageable extent. If the use of the selected digitization measures were reached, the ideas were transferred to the further measures of the "Anyway" projects or would be taken into account in future projects. One example is the above-mentioned development of standardized switchgear for digitized physical precipitation recording. In this pilot project a KOMMUNAL 4.0 idea was tested at 10 physical precipitation measuring stations. If the test run would be successful, the digitization technology of KOMMUNAL 4.0 will also be used in more than 200 measuring stations. The feasibility of more than 40 application development was checked at the project KOMMUNAL 4.0. Also corresponding application concepts and business model possibilities were examined. 20 ideas could be developed up to implementation maturity, half of which were put into practice and tested. The other half of ideas will be implemented outside of the KOMMUNAL 4.0 project starting in 2020 with the exception of four cases. This corresponds to an implementation rate of 80%. This high rate was only possible because almost all pilot projects were based on "Anyway" projects of the communal partners. Three exemplary applications are presented below.

5.2 Predicting the flooding of gullies

An exemplary example of a KOMMUNAL 4.0 pilot project is the so called "sinkbox management". It was developed and tested as one of the first ideas in close coordination with the municipal partner. All sinkbox data were already stored in the HST asset/maintenance system KANiO before the project starts. However, at the beginning it was not possible on the basis of the existing data to estimate which sink boxes were under the risk of flooding during a rainfall event, so an effective preventive maintenance with regard to future heavy rainfall event was not possible. This had to be changed by the joint project.

On the basis of 10 assessment criteria developed in cooperation with the operating people (**Figure 4**), a hazard matrix was developed that could be individually created for each sink box. The matrix was integrated into the KANiO software by connecting KOMMUNAL 4.0 platform elements. By linking the KANiO software to KOMMUNAL 4.0 platform and precipitation portal NiRA.web, an automatic data comparison of precipitation forecasts for selected urban areas with the data of the hazard matrix is now carried out. If, for example, a defined rainfall event is forecast for the selected period (e.g. >15 l/mm^2 in the next 24 hours), the data of the hazard matrix is compared with the precipitation forecast of NiRA.web and those sink boxes are identified which are most at risk. The system automatically generates a work order for the endangered sink boxes so that the affected sink boxes can be emptied and cleaned as a precaution.

5.3 Wastewater flexibility "Diemelsee 4.0"

The municipality of Diemelsee in the district of Waldeck-Frankenberg/Germany is currently constructing a new biological wastewater treatment plant by using the SBR process in the holiday resort of Heringhausen. With its 400 inhabitants, the

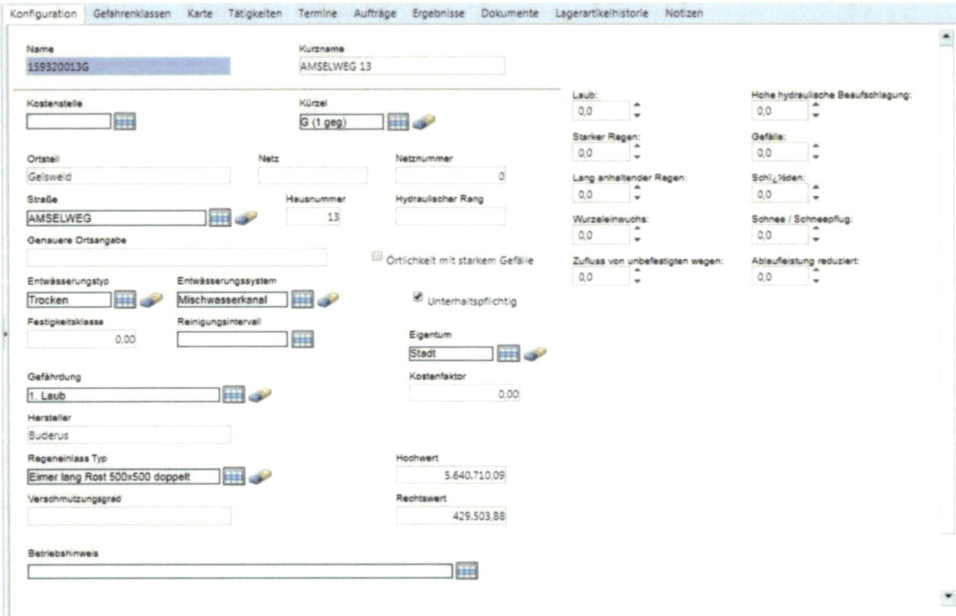

Figure 4.
Input screen sink box management (screenshot shows the input screen for one sink box with different influencing criteria like e.g. heavy rain, leaves, high hydraulic flow, street gradient, root ingrowth snow, flow from dirt roads, drainage capacity; also geodetic and type date).

town has an estimated 4000 overnight guests and 1000 day visitors in the summer months. The large number of guests leads to an extremely fluctuating amount of wastewater monthly and daily. With the help of KOMMUNAL 4.0, the idea was developed to equip the infrastructure with digital control technology to increase the flexibility and efficiency of the sewage treatment plant and the sewer network. The idea was modeled on the pilot project "Digital Sewage Plant Söllingen", which has already been reported on in detail elsewhere [11]. The wastewater treatment plant and the associated sewer network will be equipped with KOMMUNAL 4.0 control technology elements and networked with precipitation forecasts and tourism data. An additional innovative data analysis for the optimal coupling of the wastewater treatment plant with pressure pipes, pump stations, rainwater retention basins (which are connected upstream of the wastewater treatment plant), for the absorption of hydraulic peak loads and inlet fluctuations into the new SBR plant to be built and the associated sewer network are part of the project. A core element of the project is the Case-Based Reasoning (CBR) approach, which is a kind of artificial intelligence that learns from experience from previous events and derives improvements from it.

5.4 Practice-integrated learning ensures effective knowledge transfer

To ensure that even small measures from the "Anyway" projects are suitable as a start into digitization, a high level of learning and transmission success should be ensured. For this reason, KOMMUNAL 4.0 tested two further developments in practice in addition to the technical pilot projects. On the one hand it is about securing the knowledge of older employees and on the other hand it is about the question how planning, variant consideration, implementation and learning can be integrative and agilely interlinked in a common project execution. In view of the increase in municipal tasks and the simultaneous shortage of personnel and skilled workers, there is a lack of human resources to try out new developments as complementary projects. In the course of KOMMUNAL 4.0, the new methodology HELIP (Highly Efficient Learning in Projects/Processes) was developed in order to meet this challenge effectively. On the basis of current research results on learning and transfer research as well as from project management, measures such as the pilot projects presented are suitable for starting practical digitization at an early stage, even if many digitization topics still need to be learned [12]. The HELIP concept is based on a 360° reflection of the tasks and necessary learning content at the beginning of the planning phase. It assigns the necessary knowledge transfer of new contents to individual organizational contexts and the task of the respective municipality/department/division and integrates them into selected "Anything" projects. The appropriate practice-related task packages are also adapted to the further decisions and planning steps of the overall process. This ensures that the learning outcomes of smaller "Anyway" projects are optimally transferable to larger digitization projects. Learning takes place in everyday working life and is not separated from practice in remote seminars or training courses. The separation of planning/implementation and further training, which has been customary up to now, is thus abolished. In addition, HELIP supports the effective implementation of the Sustainable Development Goals No. 4, 6, 8, 11 and 13 of the United Nations and can be further developed as a basic principle for management and education in projects to achieve the goals No. 7, 9, 12, 14 and 15.

6. KOMMUNAL 4.0 ensures that infrastructures retain their value in the future

Many small and medium-sized communities are faced with the challenge of reliably planning for the future in terms of maintaining and expanding their infrastructure in view of the consequences of demographic change. It is not unusual for the largest infrastructure assets to be hidden underground. Up to 70% of this can be accounted for by the sewer system with its special structures and sewage treatment plants [13]. Sufficient and reliable data is required to achieve optimum investment planning. Decisions, based on inaccurate assumptions and estimates, must be reduced to a minimum in the future. A major role is playing a value-preserving operation of existing plants and objects, e.g. through efficient control solutions or cost-saving condition monitoring.

The basis for intelligent data management and the control and operation management is meaningful data acquisition and evaluation. This requires modern IT structures that can be used both locally and as web-based solutions. KOMMUNAL 4.0 pursues this premise and takes care of a fully comprehensive data and IT structure. This starts locally at the machines (CPS), networks the objects with each other and aims at a networked analysis and management of entire infrastructures via the web-based data and service platform. This will create a basic structure that is not limited to applications in water management alone, but will also be suitable for use in other infrastructure sectors. The start into digitization can be made from an overall strategic perspective by setting up a central data and service platform, but also on the basis of software-related or machine-related individual solutions. It must be ensured that all required individual components (see **Figure 1**) can be networked and thus integrated into the intended overall system.

Author details

Günter Müller-Czygan
HST Systemtechnik GmbH & Co. KG, Meschede, Germany

*Address all correspondence to: guenter.mueller-czygan@hst.de

References

[1] UBA. Wasserwirtschaft in Deutschland, Wasserversorgung und Abwasserbeseitigung, Stand April 2014, Zahlen für. 2010

[2] Bildquelle. Bundesverband KOMMUNAL 4.0 e.V.

[3] Uwe Frigger, Günter Müller-Czygan, Tobias Wiese. Sichere Vor-Ort-Anlagenbedienung mit mobilen Geräten, Automation Blue 2. 2017

[4] Günter Fehr, Günter Müller-Czygan, Die Digitale Kläranlage Söllingen—Pilotprojekt in KOMMUNAL 4.0, Automation Blue 4. 2017

[5] Günter Müller-Czygan, Andreas Stolz, Digitalisierung von Entlastungsschwellen, Automation Blue 1. 2018

[6] Christopher Becker, Günter Müller-Czygan, Mit smarten Maschinen in die digitale Wasserwirtschaft starten, Wasser und Abfall 9/2016

[7] EUWIDD-Interview mit Martin Weyand (BDEW) in: EUWID Wasser und Abwasser 3. 2018

[8] Available from: http://www.bmwi.de/DE/Presse/pressemitteilungen,did=720380.html

[9] Bildquelle: HST Systemtechnik GmbH & Co. KG

[10] Becker C, Müller-Czygan G. Mit smarten Maschinen in die digitale Wasserwirtschaft starten. In: Springer Vieweg, Wasser und Abfall, 9/2016. 2016

[11] Fehr G, Müller-Czygan G. Die "Digitale Kläranlage Söllingen" – Pilotprojekt in KOMMUNAL 4.0. In: Vulkan-Verlag GmbH, Automation Blue, 4-2017. 2017

[12] Müller-Czygan G.. "Empirische Untersuchung zur Bedeutung von Lerntransfer-und Umsetzungsfaktoren im Digitalisierungsvorhaben "KOMMUNAL 4.0" unter besonderer Beachtung der Wechselwirkungen von Technik und Veränderungsmanagement" (Master Thesis business psychology at FOM Hochschule für Oekonomie & Management Münster). 2018

[13] Stefan Hattenbach. Markus Vogel "Das verborgene Vermögen der Kommunen: Entwicklung einer Strategie zum Werterhalt in einer Dorfgemeinde", Tagungsband DWA Landesverbandstagung. 2015

Chapter 4

Aquifer Characterization: The Case of Hawassa City Aquifer

Shemsu Gulta and Brook Abate

Abstract

Hydrogeologists and other water experts agree on that the effective groundwater management requires: firstly, a good understanding of the aquifer system; secondly, identification of practical measures to control abstraction; and thirdly, improvement in groundwater resource through artificial recharge. A 16 years' pumping test and drilling lithology data and productive 29 wells were used to characterize the aquifer parameters of the Hawassa City, Ethiopia. The aquifer system was characterized physically, potentially, spatially, quantitatively, and qualitatively using AquiferTest software by applying Moench method to pumping test response data considering the basic assumptions in the model. Weathered and fractured pumice, basalt Scoriaceous rocks, fine-to-coarse-grained sand, and weathered ignimbrites are major water-bearing formations found from the analysis. High porosity and permeability due to these fractures are found to be a risk for the easy contamination of the ground water from surface wastes especially at the shallow aquifer water areas. Spatially, the southern corner and the lake shore of the city were identified as a huge potential area. Percentage of recovery results are 95–100% and transmissivity varies from 4.77×10^{-4} m^2/s to 1.75×10^{1} m^2/s. This follows the general pattern of increasing value from east to west, that is, the value increases from the upper part of the basin to the lower. Moreover, the annual ground flow vector map of the area was developed using static water level data to see the direction of subsurface flow in the area. Accordingly, a large magnitude of water flowing from the central and west directions to the lake shore is identified showing similar profile with the surface flow.

Keywords: Hawassa City, pumping test, AquiTest, Moench method, aquifer characterization

1. Introduction

1.1 Aquifer characterization

Characterization of hydraulic properties of an aquifer involves the use of existing pumping test data, hydrogeological map, geologic map, soil map, lithology obtained from well logs, aquifer thickness, water table depth, structures and surface water features, etc., so as to analyze the lateral distribution and nature of the aquifer.

Pumping test analysis and well logs were used to identify the aquifer system of Tarmaber formation by the author [1]. Accordingly, the formation is categorized as fractured aquifer where the dominant aquifer types are confined-double porosity and single plane vertical aquifer. The double porosity aquifers are related to deeply drilled wells reflecting presence of large and narrow fracture systems with high permeability but with lower storage capacity. Transmissivity varies between 0.5 and 1400 m^2/day.

1.2 Pumping test, slug test, and recovery measurement

1.2.1 Pumping test

Aquifer pumping test involves posing artificial stress on the hydrogeological system by pumping water from a well and measuring the changes in water levels in the pumped well and nearby observation wells. The response of the hydraulic head in the aquifer can be used to estimate transmissivity or hydraulic conductivity in the particular aquifer. The data from the pumping tests are used to calculate the specific capacity (specific capacity = Q/s, where Q = yield and s = drawdown) of the well.

The total drawdown is the algebraic sum of the individual drawdowns during the constant rate of pumping. The total drawdown can be estimated as:

$$TDD_W = DWL - SWL \tag{1}$$

where TDD_W is total drawdown, DWL is dynamic water level, and SWL is static water level.

In addition to estimating hydraulic properties of an aquifer system such as transmissivity and hydraulic conductivity, a step-drawdown test is made to evaluate well performance criteria such as well loss and well efficiency. All conventional well hydraulic theories are based on the assumption that laminar flow conditions prevail in the aquifer during pumping.

In our study area Lake Hawassa catchment, pumping tests are most commonly conducted following the development works usually for 24 h. The objective of the pumping test is to evaluate the production capacity of the well and the aquifers and to decide on the capacity and the position of the production well. Moreover, the results of the pumping test are of great value to be used in the well operation and maintenance apart from being used for future ground water study and research purposes. A pumping well with a pump that allows to control the pumping rate, one or more observation wells close enough to see the influence of the pumping well (mostly in the pumped well or control well itself), some means of measuring water levels in the observation wells at specific times throughout the course of the test; using automatic pressure transducers or manually using depth meters.

1.2.2 Slug tests

Multiple well pump tests are not always feasible—there may not be any observation wells, and it may cost too much to put new wells in or it could be that we are dealing with a contaminated system, and everything we pump out of the well have to be hauled away and treated, which need more investment. There are alternative methods that involve piezometers and a general set of tests called "slug" tests. These tests involve introducing or removing a known quantity of water (a "slug" of water) from a piezometer and measuring the time it takes to recover to the initial static water level [2].

1.2.3 Recovery measurement

After a pump test is performed for the specified time period, the well is shut off and recovered data (heads in the observation well at times after the pump is shut off) are collected. The recovery is the rate at which water in the well returns to its static water level after the pumping is turned off. The recovery measure is another estimate of the well yield. Percentage of recovery can be estimated by the following relationship:

$$PR = 100 \times (DWL - MRWL)/TDD_W \tag{2}$$

where PR is percentage recovery, DWL is dynamic water level, MRWL is the maximum recovery of water level, and TDD_W is total drawdown.

1.2.4 Specific capacity

Specific capacity is known as well performance test estimated by assuming the well is pumped at a constant rate long enough to establish the equilibrium drawdown; within the well, there is a combination of the decrease in hydraulic head (pressure) and a pressure loss due to well loss. Specific capacity is defined as:

$$Sc = Q\,/TDD_w \tag{3}$$

where Sc is specific capacity, Q is the pumping rate, and TDD_w is the total drawdown in the well due to both aquifer drawdown and well loss.

Well loss is created by the turbulent flow of water through the well screen and into the pump intake. Specific capacity is estimated by discharge on a linear x-axis and drawdown on a linear y-axis and measuring the slope of the straight line fit.

1.2.5 Analyzing and evaluating pumping test data

Different hydrogeologists [2, 3] suggested very similar procedures for analyzing and evaluating pumping test data. In general, it is as much an art as a science. It is a science because it is based on theoretical models that the hydrogeologist must understand and thorough investigations must be conducted into the geological formations in the area of the test. It is an art because different types of aquifers can exhibit similar drawdown behaviors, which demand interpretational skills on the part of the hydrogeologist.

During an aquifer test, the hydraulic head in the aquifer declines as the time of pumping increases. Analysis of hydraulic head decline, or drawdown, allows for the estimation of aquifer hydraulic properties. For instance, authors [4] in the study of aquifer parameters estimation in basaltic terrain and the application of wireless sensor networks at Chikhaldara region, India; identified a pumping test as the best available method to evaluate aquifer parameters. The tests were performed at 20 locations using the local farmers' well pumps. The pumping phase of the tests had a short duration of 60–210 min; the recovery phase of the tests had a longer duration of 90–300 min. Three methods were adapted to estimate the aquifer parameters in a basaltic terrain. Out of the three methods, two were conventional or analytical curve matching techniques (the study is found in [5, 6]). The other technique was a numerical method. Moreover, this study determined the flow direction of sub-surface water using static groundwater level data within a basin from the past 20 years (1972–1992); an annual average water level map was constructed (with respect to the above mean sea level).

1.2.6 Pumping tests and methods of analysis

Among the main techniques are analytical/conventional methods and numerical methods. Analytical/conventional methods involve one of the curve matching, finding inflection points, or for special cases and fitting straight lines to the pumping test data [4].

Models for the interpretation of pumping test data were initiated under constant pumping test rate and equilibrium conditions for confined and unconfined aquifers. Since then, different methods have been designed for pumping test analysis. Reliable estimates of the hydraulic parameters controlling an unconfined aquifer's capacity to store and transmit water are generally obtained by pumping test analysis with one or more analytical models, of which authors [5, 7–14] are the most popular. Nowadays, the entire computation procedures and hydrological equations are typically written into computer programs.

However, each of the methods is based on basic assumptions relating to geologic formation, the basic types of well, such as well diameter, dug well and bore well. Therefore, it is important to choose the right method of interpretation based on the field conditions [3].

The study by Mishra and Kuhlman [6] discussed concisely the issue of which model for which well conditions. According to this study, analytic and semi-analytic solutions are often used by researchers and practitioners to estimate aquifer parameters from unconfined aquifer pumping tests. The nonlinearity associated with unconfined (i.e., water table) aquifer tests make their analysis more complex than confined tests.

As the method by Cooper and Jacob [5] is a simplification of the Theis method solution, the pumping well should fully penetrate a confined, homogeneous, and isotropic aquifer. Single well tests from partially penetrating wells in unconfined aquifers depart greatly from the Theis model. Moreover, unconfined aquifer tests are affected by vertical anisotropy and specific yield in addition to transmissivity and storage coefficient. These additional parameters control vertical gradients that are created by partial penetration and drainage from the water table. Likewise, leakage from adjacent confining beds also could affect transmissivity estimates, which likely will be overestimated by the Cooper-Jacob method [15].

The study in Neuman [7] presented a physically based mathematical model that treated the unconfined aquifer as compressible and the water table as a moving material boundary. Newman's approach describes the aquifer delayed response was caused by physical water table movement; therefore, it was proposed to replace the phrase "delayed yield" by "delayed water table response." Besides this, the model exhibits three distinct drawdown segments as shown in **Figure 1**.

Early-time response is controlled by the transmissivity and elastic storage coefficient and is analogous to the response of a confined aquifer, and the water table

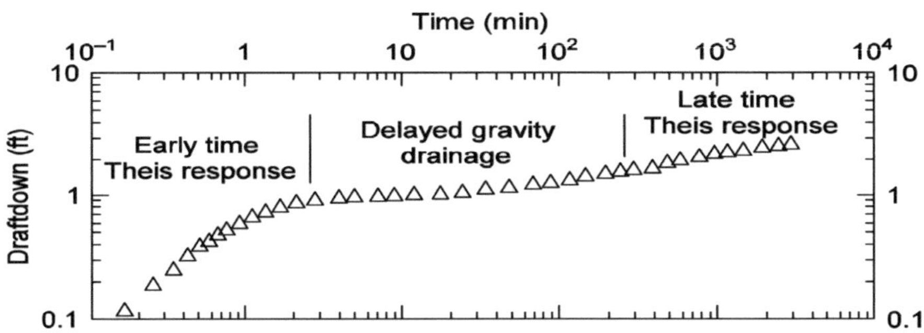

Figure 1.
The three distinct drawdown segments in an unconfined aquifer (from [16]).

does not drop significantly. Late-time response is a function of transmissivity and specific yield (drainable porosity). Release of water is due to drainage from formation over large area water table decline slows and flow is essentially horizontal. At intermediate time, the response is controlled by the aquifer's vertical hydraulic conductivity. The release of water is from gravity drainage and slope of time-drawdown curve relative to Theis curve decreases.

After comparative analysis of various methods for determination of specific yield, the author in [8] concluded that the water table response to pumping is a much faster phenomenon than drainage in the unsaturated zone above it.

The analytical model developed by Moench [14] combines and extends the work of Boulton [13] and of Neuman [7, 8] to account for the release of water from the unsaturated zone above the water table. In spite of the possible limitations inherent in the assumption, the Neuman model has been used successfully by many hydrogeologists in analysis of pumping tests conducted in water table (unconfined aquifers). One of the primary features of the model is that it allows drawdown to vary continuously in the vertical as well as the horizontal directions, thus retaining the full three-dimensional, axisymmetric character of the flow regime. Another feature is that the model accounts for aquifer compressibility. However, the use of Neuman type curve fitting for unconfined aquifer conditions has sometimes led to values of specific yield that are unrealistically low (plus sometime too high) compared to volume-based calculations.

Moreover, although both the Boulton and Neuman models could account for the compressive characteristics of an aquifer by assuming the pumped well is infinitesimal in diameter, it becomes impossible to account for effects of well bore storage, thereby limiting the usefulness of the models for accurate evaluation of specific storage. This assumption necessitates that observation wells be located at large distances from the pumped well to reduce the influence of well bore storage. Unfortunately, this last requirement makes it difficult to record accurate early-time measurements due to small drawdowns at large distances.

The study by Neuman [7] attributed the inability of Neuman's models to give reasonable estimates of specific yield (S_y) and capture this observed behavior near the water table due to the disregard of "gradual drainage." To resolve this problem, the instantaneous moving water table boundary condition used by Neuman was replaced with one containing a Boulton [13] delayed yield convolution integral. The study by Neuman [7] recommended the composite analysis of pumping test data and grouping of corresponding time drawdown data for parameterization as opposed to the analyses of individual drawdown curves.

The Moench solution, presented in AquiferTest V. 2.55, is an extension of the Neuman solution for drawdown in a homogeneous anisotropic confined or unconfined aquifer with fully or partially penetrating pumping and multiple observation wells. The Moench solution also allows for water in the overlying unsaturated zone to be released either instantaneously in response to a declining water table or gradually as approximated by Boulton's convolution integral.

1.2.7 AquiferTest software application for pumping test data analysis

There are a number of software programs that can be used to complete the data analysis of aquifer test drawdown data that include, but are not limited to, AQTESOLV, AquiferTest, WTAQ, and AquiferWin32.

In the analysis of a multiple pumping test conducted in a layered unconfined aquifer (harbor area of Antwerp, Belgium), the use of two computer programs was presented: AquiferTest and WTAQ to investigate and compare previous results obtained for transmissivity, hydraulic conductivity specific yield, and storage coefficient. The study made use of the Theis-type [14] curve method in AquiferTest applicable to both partially and fully penetrating wells. This was used to calculate dimensional drawdowns that are compared with time-drawdown data from 23 observation points to estimate the hydraulic properties of a finite, layered unconfined aquifer situated in the harbor area of Antwerp. The study concluded that AquiferTest and WTAQ form an excellent pair for the analyses of single or multiple pumping tests in unconfined aquifers.

An assessment was made for the hydraulic properties of the Ethiopian Ashange formations applying AquiferTest software. In the study, a total of 70 wells raw pumping test data were analyzed and used besides their respective lithological log to determine hydraulic property of Ashange formation. This study has done identification, analysis, and interpretation of aquifer system hydraulic properties of the geologic formation using the secondary well pump test data, lithological log, and data of hydro geological field observations. Among the different stages of pumping tests, constant rate pumping tests lasting between 5 and 72 h and recovery tests were used to determine transmissivity, hydraulic conductivity, and storativity values. The study analyzed single pumping test data mainly using Theis time-drawdown graphic method by which aquifer properties have been calculated. The pump test data including measured and calculated ones have been organized and processed using the Aquifer test software version 3.5. Arc GIS 9.2 and Global mapper 11 were also used for mapping in that study. As a result, the study finally identified the aquifer characteristics of the Ashange formation with respect to depth of the boreholes, age, and variation of its spatial distribution and groundwater potential.

2. Geological and hydrogeological setting

In this phase of the research work, the aquifer type and water-bearing formations of borehole sites are generalized on the basis of lithological logs developed from well cutting logs collected. For the study, 29 boreholes were used to understand the Hawassa City aquifer.

The main water-bearing geological formation in the Hawassa City ground water system is classified along with the respective water-bearing thickness. The northeast part of the city has water-bearing formation of pumice ash and sand at depth of 14–20 m, weathered and fractured basalt formations at a depth 23–40 m, weathered pumice and rhyolite (42–54 m), and black and red scoria (60–66 m).

The north part of the city has sand (25.52–29 m), trachyte (29–39 m), and the volcanic origins' rocks (39–42 m) and (39–50.5 m). This shows that the area is likely dominated by volcanic rocks for depth below 39 m.

The lake shore (east and southeast of the Hawassa city) covers slightly rhyolite (7.15–9 m), scoria (9–12 m), dominant weathered basalt (12–27 m) and course sand scoriaceous basalt that covers up to 39 m depth of the area. As the well site goes apart to the west direction, the hydrogeology appears different for the whole lake shore; the water striking depth is increasing, and ash with sand formation (23.58–31 m), fractured basalt (49.43–57 m), and ash with scoriaceous basalt (60–69.24 m) are recurrently reported. The far southern part of Hawassa city (*Gara Riqata area*) where relatively the deepest wells of this study are located, the water striking point gets deeper and the major aquifers recognized are pumice type of fractured, weathered and course-grained (18–32 m), highly weathered pumice (32–58 m), fine-to--medium-grained sand (66–72 m), silty sand and weathered rhyolite (79–94 m), fractured pumice (94–102 m), weathered pumice (102–120 m), fine-to-medium-grained sand (120–166 m), weathered pumice (166–172 m) and fine-grained sand (184–196 m) are dominant of which sand covers the largest formation.

Around the western part of the city (the industry zone), the water striking point is the deepest of the study area. Highly fractured and weathered scoriaceous formation dominates the water-bearing strata (52–84 m). The central areas generally fractured basalt (12–21.5 m), sand and ignimbrite (22.64–33.54 m), scoria and pumice (27.38–38.20 m), and highly weathered ignimbrite (39.27–45 m).

About 30–60 m ignimbrites and pumice are dominant in large area of the central part. These ignimbrite and pumice of the rift floor are well jointed while in some cases, it is massive and pumiceous. Where it is well jointed, it has a high or moderate permeability, but in the other part, it has low permeability.

The relationship between lithology and aquifer characteristics is used to understand the qualitative and quantitative aspects of the hydrogeology in these areas. The study by Glenn and Duffield [17] established the estimate of the representative range of hydraulic properties (horizontal and vertical hydraulic conductivity, storativity, specific yield, and porosity) of aquifers and aquitards in relation to the formation type using values reported in different literatures. These tabulated values are used to understand the hydraulic properties of the study area.

Therefore, the dominant water-bearing formations (weathered pumice, scoria, fractured basalt, and sand of different types) possess large pores. Pumice and fractured basalts strata, which are common relatively in the shallower formations, are devoid of primary openings but possess secondary openings in the form of fractures and joints. These features aid in the infiltration of surface water. Besides, pores and fractures in laterites and fractures and joints in basalts act as reservoirs of groundwater.

Highly fractured and weathered scoriaceous formation dominates the water-bearing strata (52–84 m), and the fine-to-course-grained sand is the main water source in depth beyond 100 m. Furthermore, lack of confining rocks like clay in the area studied indicates that groundwater occurs in phreatic, unconfined conditions in the weathered basalts that outcrop at the surface.

Looking into representative values, aquifers in the area are high hydraulic conductivity units and large porosity which will produce higher and more sustained well yields than an aquifer where the clean sands and gravels are compartmentalized by interbedding with clay and other low hydraulic conductivity units.

2.1 Aquifer physical properties of Hawassa City

The results show (**Table 1**) the depth ranges from 25 m to 200 m below the surface. The pumping phase of the tests had a duration of 1440 min; the recovery phase of the tests had a duration of 45–240 min. Constant rate of discharge was applied for each of the wells. These constant discharge rates are from 3.0 l/s (for Gara Riqita 6) to 66 l/s (for Gara Riqita 1). Total drawdown varied from 0.03 m (for Zewdu Village) to 12.36 m (for HU Techno Village) and average of 2.53 m.

2.1.1 Specific capacity

The specific capacity of the wells as the ratio of the yield to the total drawdown is determined using Eq. (1). These two parameters (TDDw and Sc) along with the discharge rate are calculated and tabulated for 29 wells as shown in **Table 1**.

As per **Table 1**, values of specific capacity range from 0.54 l/s/m to 2200 l/s/m. Maximum values are toward the southwest part of the city (at Hawassa University Referral Hospital) and tend to decline toward the central and then to the northern corner. A decline in specific capacity may indicate declining S or T values due to declining water levels or piezometric surfaces, thus large water level drawdown for the specified discharge rate. It can also be used to determine the distribution of transmissivity in the aquifer. The spatial distribution of specific capacity reveals

Well name	Depth of BH(m)	B (m)	DWL (m)	SWL (m)	TDDW (m)	Q (l/sec)	Sc (l/sec/m)
SNNP council	42	8.00	38.17	34.60	3.57	4.45	1.25
Zewd Village	25	14.15	25.61	5.60	0.03	3.66	122.00
South star Int. H.	52	26	25.61	25.60	0.01	6.50	650.00
Agrostone factory	86	11.40	24.54	18.50	6.04	22.00	3.64
S.police garage	52	25.00	30.30	20.20	10.10	6.00	0.59
Abebe W.private C	41.5	17.00	7.27	5.63	1.64	61.00	37.20
Gara.R1	200	98.00	4.81	2.87	1.94	66.00	34.02
Gara.R4	193	30.00	10.76	10.60	0.16	63.60	397.50
Gara.R5	186	64. 00	14.79	4.46	10.33	64.80	6.27
Gara.R6	168	136.00	27.74	27.72	0.02	3.00	150.00
Zinabu Abera	41.5	11.32	52.13	52.12	0.01	5.00	500.00
Dairy farm	67	14.15	22.30	22.20	0.10	5.00	50.00
South Roads Auto.	37	11.32	46.50	45.70	0.80	7.33	9.16
Moha soft drink	90	70.00	25.67	25.30	0.37	5.24	14.16
Midroc con.	52	17.04	17.82	17.20	0.62	7.00	11.29
Awassa agi.r.cenr	50	30.00	56.73	56.72	0.01	5.00	500.00
TTC	71.5	19.81	25.46	25.39	0.07	20.00	285.71
HU 1 (White house)	86	36.00	27.65	27.50	0.15	22.00	146.67
HU 2 (Techno library)	50	30.00	42.53	30.17	12.36	20.00	1.62
HU3 (IoT village)	58	24.00	29.71	24.80	4.91	20.00	4.07
HU 4 (Staff complex)	50	18.00	30.08	28.00	2.08	22.00	10.58
HU 5 (Green house)	46	18.00	12.10	12.02	0.08	22.00	275.00
Dashen bank	36	16.00	32.10	26.70	5.40	7.00	1.30
ATENET S.C	48	18.00	17.50	16.40	1.10	20.00	18.18
HU Condominium	51	24.00	7.70	7.60	0.10	22.00	220.00
ACA HU	40	18.20	13.55	13.54	0.01	22.00	2200.00
HU (Health sc college)	41	23.50	23.94	15.54	8.40	6.50	0.77
SOS Village	60	32.00	16.66	13.72	2.94	6.66	2.27

Table 1.
Discharge rate, total drawdown, and specific capacity results.

that the increase in values coincides with the storage coefficient or transmissivity value presented in **Table 1**. Higher specific capacity values were also found to coincide with areas where extension fracture systems occur.

Using the total drawdown of the wells, a contour map is developed (**Figure 2**) to see the response of the wells at the end of the pump test duration. This is important to conclude about the potential of the aquifer for discharging.

This map shows, at the final hours, the water level that has been nearly stabilized at the end at about water strike zone for those high potential wells of smaller drawdown. This fluctuation could be due to the difference in the rock type of that

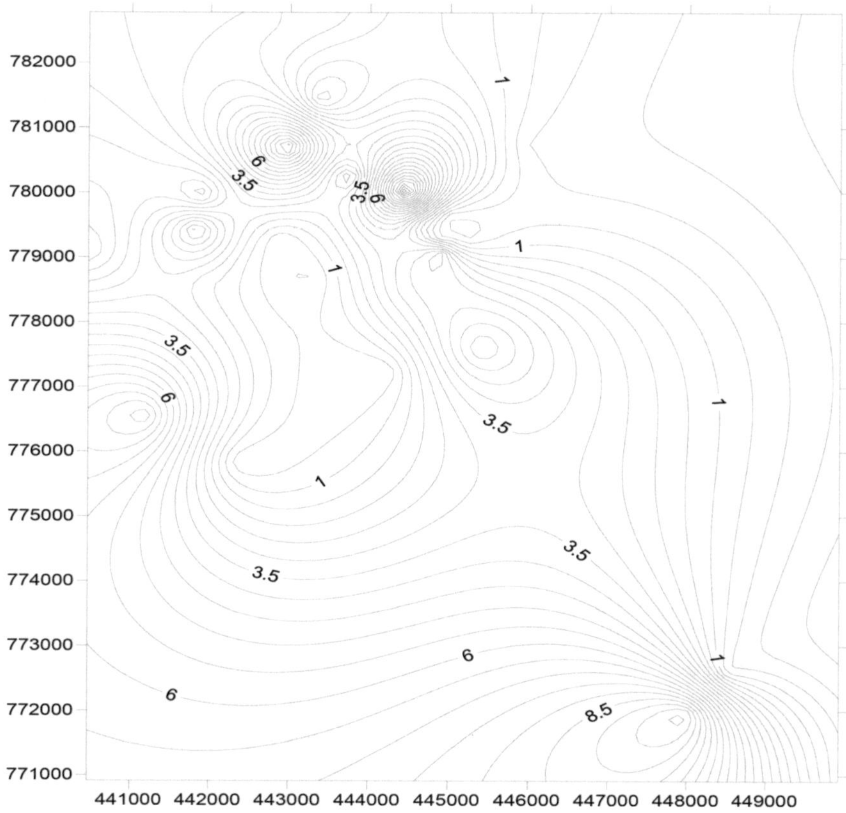

Figure 2.
Contour map of the total drawdowns (UTM Coordinates at Zone 37).

area. Percentage of recovery is also estimated using TDDw and the maximum recovery of water level during pumping test using Eq. (3). The results have 95–100% recovery percentage which reflects high recovery rate. The recovery of these wells is also fast that they recover up to the SWL within few minutes, even in 30 s. The results clearly show that the study area is of high potential of ground water with fast recovery for dewatering of the aquifer.

2.1.2 Saturated aquifer thickness (b)

The thickness values found in the study area range from 8 m to 136 m and average of 30.30 m. Within the study area, the spatial distribution of saturated thickness clearly indicates that the depth and the lakeshore have significant effect on it. As the cutting depth increases and the site getting nearer to the lake shore, the water-bearing formation thickness increases.

2.1.3 Analysis results using Moench method

This hydrogeological analysis method is done to determine the important hydrogeological parameters of the wells using the method selected. Along with the curve, the analysis result displays the important parameters, that is, transmissivity, hydraulic conductivity, aquifer thickness, and storativity (specific yield for unconfined aquifers). The result of the hydraulic parameters determined for the analysis of all the wells are tabulated in **Table 2**.

As presented in **Table 2**, the three parameters are discussed below.

2.1.3.1 Specific yield and storativity

If the aquifer is considered as a confined one, the storativity is determined, and if it is unconfined, its specific yield is determined. The results show that confined aquifers have very low storativity values (much less than 0.01, and as little as 10^{-5}), which mean that the aquifer is storing water using the mechanisms of aquifer matrix expansion and the compressibility of water, which typically are both quite small quantities. Unconfined aquifers have specific yield greater than 0.01 which is about 0.517 for the Hawassa City subsurface.

As per the results, the specific yield is high in the south, west, central, and southwest lake shore parts of the area and low in the east and northeast corner part of the area. The storativity or/and specific yield values generally range from 4.77×10^{-4} to 5.17×10^{-1}. Further, it could be seen that there is a decrease of specific yield in the eastern to southeastern parts and again an increase toward the west and central areas.

2.1.3.2 Transmissivity

Results show that the value of transmissivity varies from 4.77×10^{-4} m^2/s to 1.75×10^{1} m^2/s. This follows the general pattern of increasing value from east to west (the lake shore), that is, the value increases from the upper part of the basin to the lower. This also shows a gradual increase of the hydraulic gradient. The high transmissivity coincides with areas where the fractured zone occurs.

This high transmissivity in the study area is a better indicator of the water production capacity of an aquifer than hydraulic conductivity. To see why, if we consider a thin aquifer, for example, a sand bed interbedded (sandwiched) between thick clay layers. The bed has a very high hydraulic conductivity because it consists of clean sand; however, if it is not thick, it will not sustain a large production well (its transmissivity is low).

2.2 Ground water flow dynamics

Annual ground water flow of the Hawassa City aquifer system is determined using wells that are tested at same year and similar season of that year. This is aimed to understand the spatial and temporal regional ground water flow pattern which is essential for managing local and regional groundwater resources, protecting groundwater quality, and delineating wellhead protection zones or drinking water supply source areas. To develop hydraulic head distributions contour map of the area, SWL and DWL data were used from **Table 3**.

From **Table 3**, five batches of wells are selected since their pumping test and completion are undertaken at nearly similar season of that year. In this respect, for 2006, 4 wells; for 2009, 6 wells, for 2012, 5 wells; for 2013, 4 wells; and for 2014, 4 wells are annually grouped along with their SWL and DWL data.

To understand the spatial trend of the flow, contour maps and vector maps are developed (**Figure 3A–J**). During vector map development, the vector orientation is reversed using the command in the Surfer software. Because the SWL and DWL readings are from the top surface downward and the software assumes the values as elevation points otherwise and then wrong flow direction will be identified.

From the above five sets of graphs (**Figure 3A–J**), groundwater moves from higher elevations to lower elevations and from locations of higher pressure to locations of lower pressure.

Well Name	**B (m)**	**T (m^2/s)**	**Kh (m/s)**	**Kv (m/s)**	**S**	**Sy**
SNNP council	8.00	3.5×10^{-3}	4.37×10^{-4}	4.37×10^{-5}	3.5×10^{-3}	
Zewd Village	14.15	2.94×10^{-2}	2.08×10^{-3}	2.08×10^{-4}		94×10^{-2}
South star Int. H.	26.00	1.75×10^{1}	6.73×10^{-1}	6.73×10^{-2}	1.75×10^{-2}	
Agrostone factory	11.40	5.17×10^{-1}	4.53×10^{-2}	4.53×10^{-3}		5.17×10^{-1}
S.police garage	25.00	1.75×10^{-2}	7.00×10^{-4}	7.00×10^{-5}		1.75×10^{-2}
Abebe W.private C	17.00	4.77×10^{-4}	2.80×10^{-2}	2.80×10^{-3}	4.77×10^{-4}	
Gara.R1	98.00	4.85×10^{-2}	4.95×10^{-4}	4.95×10^{-5}		4.85×10^{-2}
Gara.R4	30.00	5.25×10^{-2}	1.75×10^{-3}	1.75×10^{-4}		5.25×10^{-2}
Gara.R5	64.00	5.06×10^{-1}	7.90×10^{-3}	7.90×10^{-4}		5.06×10^{-1}
Gara.R6	136.00	5.15×10^{-3}	3.79×10^{-5}	3.79×10^{-6}	5.15×10^{-3}	
Zinabu Abera	11.32	2.38×10^{-1}	2.10×10^{-2}	2.10×10^{-3}		2.38×10^{-1}
Dairy farm	14.15	1.75×10^{-3}	7.44×10^{-5}	7.44×10^{-6}	1.75×10^{-3}	
Moha soft drink	70.00	5.80×10^{-2}	8.29×10^{-4}	8.29×10^{-5}		5.80×10^{-2}
Midroc con.	17.04	1.59×10^{-3}	9.39×10^{-5}	9.39×10^{-6}	1.59×10^{-3}	
Awassa agi.r.cenr	30.00	5.57×10^{-2}	1.85×10^{-3}	1.85×10^{-4}		5.57×10^{-2}
HU 1 (White house)	36.00	1.59×10^{0}	4.42×10^{-2}	4.42×10^{-3}	1.59×10^{-3}	
HU 2 (Techno library)	30.00	1.75×10^{-1}	5.83×10^{-3}	5.83×10^{-4}		1.75×10^{-1}
HU3 (IoT village)	24.00	1.59×10^{-3}	6.63×10^{-5}	6.63×10^{-6}	1.59×10^{-3}	
HU 4 (Staff complex)	18.00	1.59×10^{-2}	8.84×10^{-4}	8.84×10^{-5}		1.59×10^{-2}
HU 5 (Green house)	18.00	1.75×10^{-2}	9.72×10^{-4}	9.72×10^{-5}		1.75×10^{-2}
Dashen bank	16.00	1.75×10^{0}	1.09×10^{-1}	1.09×10^{-2}	1.75×10^{-3}	
ATENET S.C	18. 00	5.57×10^{-3}	3.09×10^{-4}	3.09×10^{-5}	5.57×10^{-3}	
HU Condom	24.00	1.59×10^{-2}	5.62×10^{-4}	5.62×10^{-5}		1.59×10^{-2}
ACA HU	18.20	1.75×10^{0}	6.25×10^{-2}	6.25×10^{-3}	1.75×10^{-3}	
HU (Health sc college)	23.50	1.75×10^{0}	7.44×10^{-2}	7.44×10^{-3}	1.75×10^{-3}	
SOS Village	32.00	5.17×10^{-3}	1.72×10^{-4}	1.72×10^{-5}	5.17×10^{-3}	
Awassa Flour	20.50	5.33×10^{-3}	2.59×10^{-4}	2.59×10^{-5}	5.33×10^{-3}	

Table 2.
Analysis result of the hydraulic parameters in Hawassa City.

Borehole name	Pumping test time D/M/Y	UTM Esting(m)	UTM Northing(m)	Depth of Well(m)	DWL (m)	SWL (m)	Q (l/sec)
SNNP council	21/04/2011	444359	778478	42.00	38.17	34.60	4.45
Zewd Village	31/12/2007	440465	779086	25.00	25.61	5.60	3.66
South star Int. H.	19/07/2010	442876	779278	52.00	25.61	25.60	6.50
Agrostone factory	12/12/2009	445387	777593	86.00	24.54	18.50	22.00
S.police garage	27/03/2010	442971	780703	52.00	30.30	20.20	6.00
Abebe W.private C	05/09/2009	441936	778609	41.50	7.27	5.63	61.00
Gara.R1	13/08/2012	449214	771791	200.00	4.81	2.87	66.00
Gara.R4	30/06/2012	449926	770917	193.00	10.76	10.60	63.60
Gara.R5	19/07/2012	447977	771948	186.00	14.79	4.46	64.80
Gara.R6	13/07/2012	448528	772643	168.00	27.74	27.72	3.00
Zinabu Abera	30/05/2006	443296	778694	41.50	52.13	52.12	5.00
Dairy farm	04/06/2006	442354	775842	67.00	22.30	22.20	5.00
Moha soft drink	28/05/2006	443745	780233	37.00	46.50	45.70	7.33
Midroc con.	24/05/2009	448915	777121	90.00	25.67	25.30	5.24
Awassa agi.r.cenr	21/12/2007	444865	779285	52.00	17.82	17.20	7.00
HU 1 (White house)	05/11/1998	445957	780765	50.00	56.73	56.72	5.00
HU 2 (Techno library)	03/05/2006	444224	777249	71.50	25.46	25.39	20.00
HU3 (IoT village)	30/04/2013	444975	779643	86.00	27.65	27.50	22.00
HU 4 (Staff complex)	12/06/2013	444444	780020	50.00	42.53	30.17	20.00
HU 5 (Green house)	08/04/2013	444868	779078	58.00	29.71	24.80	20.00
Dashen bank	15/04/2013	444620	779503	50.00	30.08	28.00	22.00
ATENET S.C	11/04/2014	445205	779257	46.00	12.10	12.02	22.00
HU Condom	26/08/2009	441827	779428	36.00	32.10	26.70	7.00
ACA HU	20/06/2009	442431	778944	48.00	17.50	16.40	20.00
HU (Health sc college)	25/03/2014	443374	781427	51.00	7.70	7.60	22.00
SOS Village	28/03/2014	441886	779967	40.00	13.55	13.54	22.00
Awassa Flour	30/03/2014	441190	776556	41.00	23.94	15.54	6.50
SNNP council	08/07/2009	442919	782411	60.00	16.66	13.72	6.66
Zewd Village	19/05/2012	443451	782781	50.00	16.66	13.72	6.66

Table 3.
The ground water level data with space and time.

In the vector maps shown, the arrow symbol points in the "downhill" direction of water table and the length of the arrow depends on the magnitude, or steepness, of the slope. A vector is drawn at each grid node; however, some nodes are skipped by changing the frequency setting for the better view of the contour. Since the grid contains dynamic water level data of wells, the direction arrows point in the direction of water flows—from high water elevation to low water elevation. Magnitude is indicated by arrow length. Therefore, the steeper slopes would have longer arrows.

Looking into the map of the study area, SWL and DWL vectors indicate similar trends for each of the years. The results show that for the year 2006, medium to high magnitude of water flowed from the northern and central parts to west and S-W direction. This indicates that there was high discharge from the western and

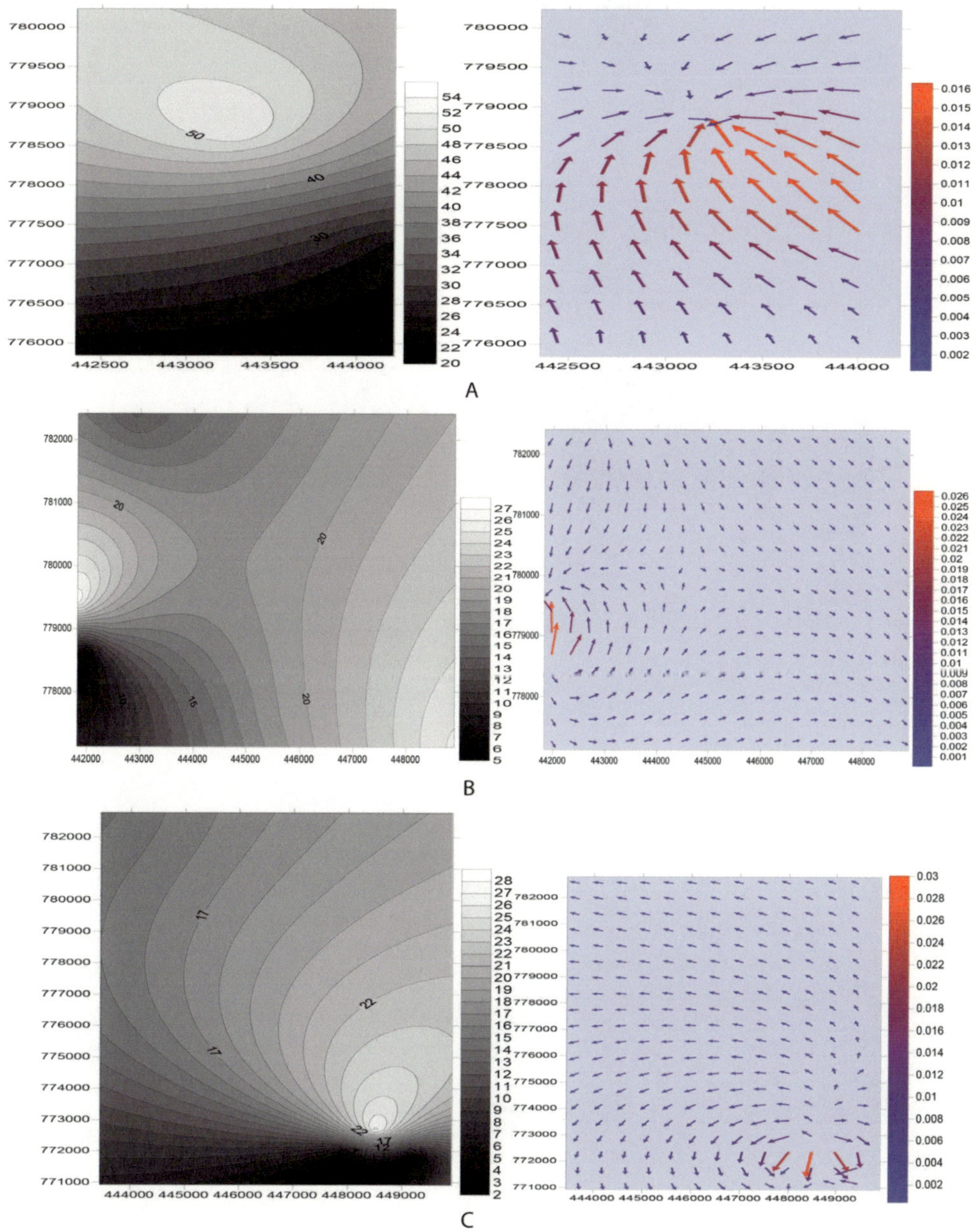

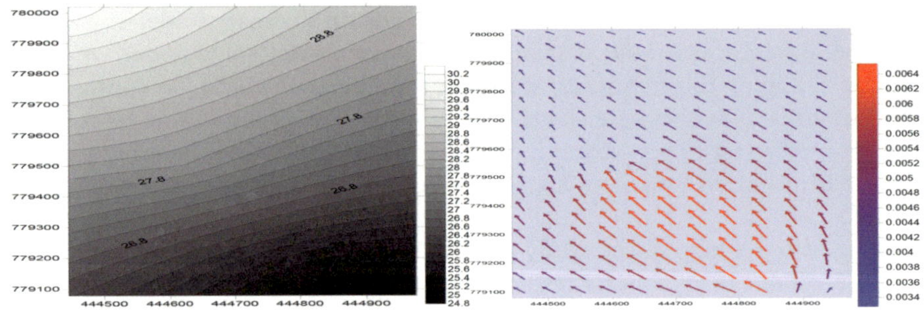
780000
779900
779800
779700
779600
779500
779400
779300
779200
779100
444500
444600
444700
444800
444900
27.8
26.8
28.8
30.2
24.8
0.0064
0.0034

D

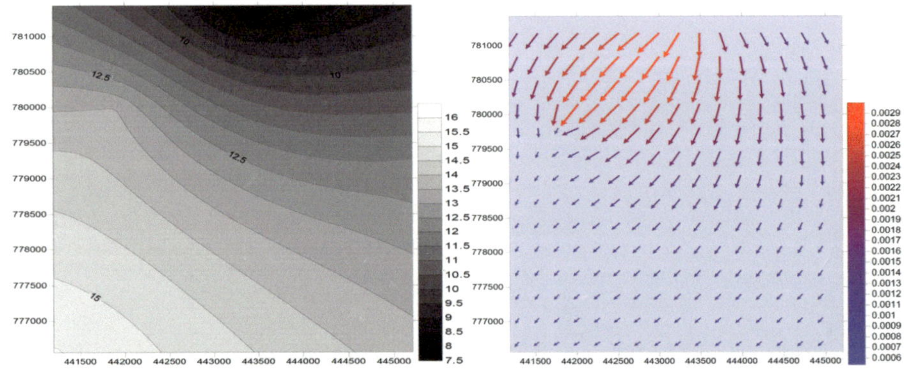
781000
780500
780000
779500
779000
778500
778000
777500
777000
441500
442000
442500
443000
443500
444000
444500
445000
10
12.5
15
16
7.5
0.0029
0.0006

E

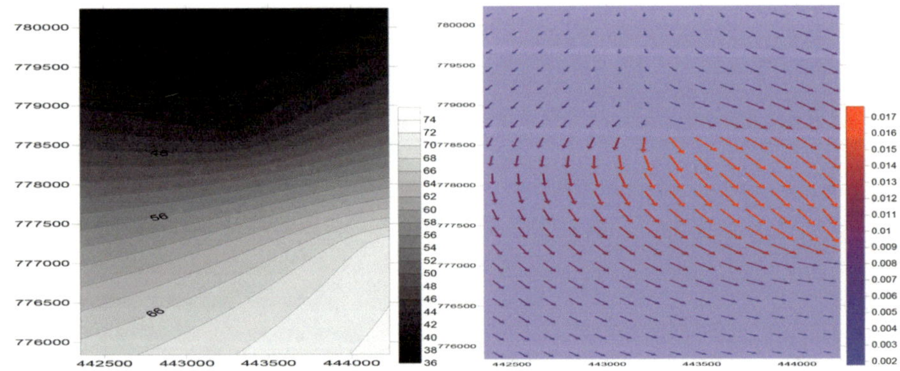
780000
779500
779000
778500
778000
777500
777000
776500
776000
442500
443000
443500
444000
56
66
74
36
0.017
0.002

F

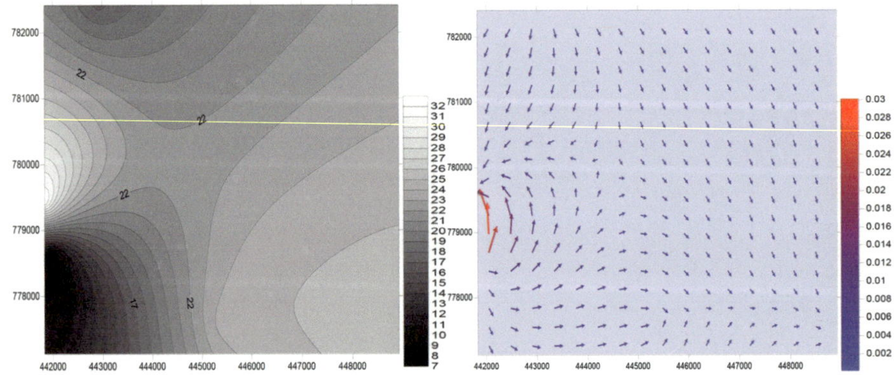
782000
781000
780000
779000
778000
442000
443000
444000
445000
446000
447000
448000
22
17
32
7
0.03
0.002

G

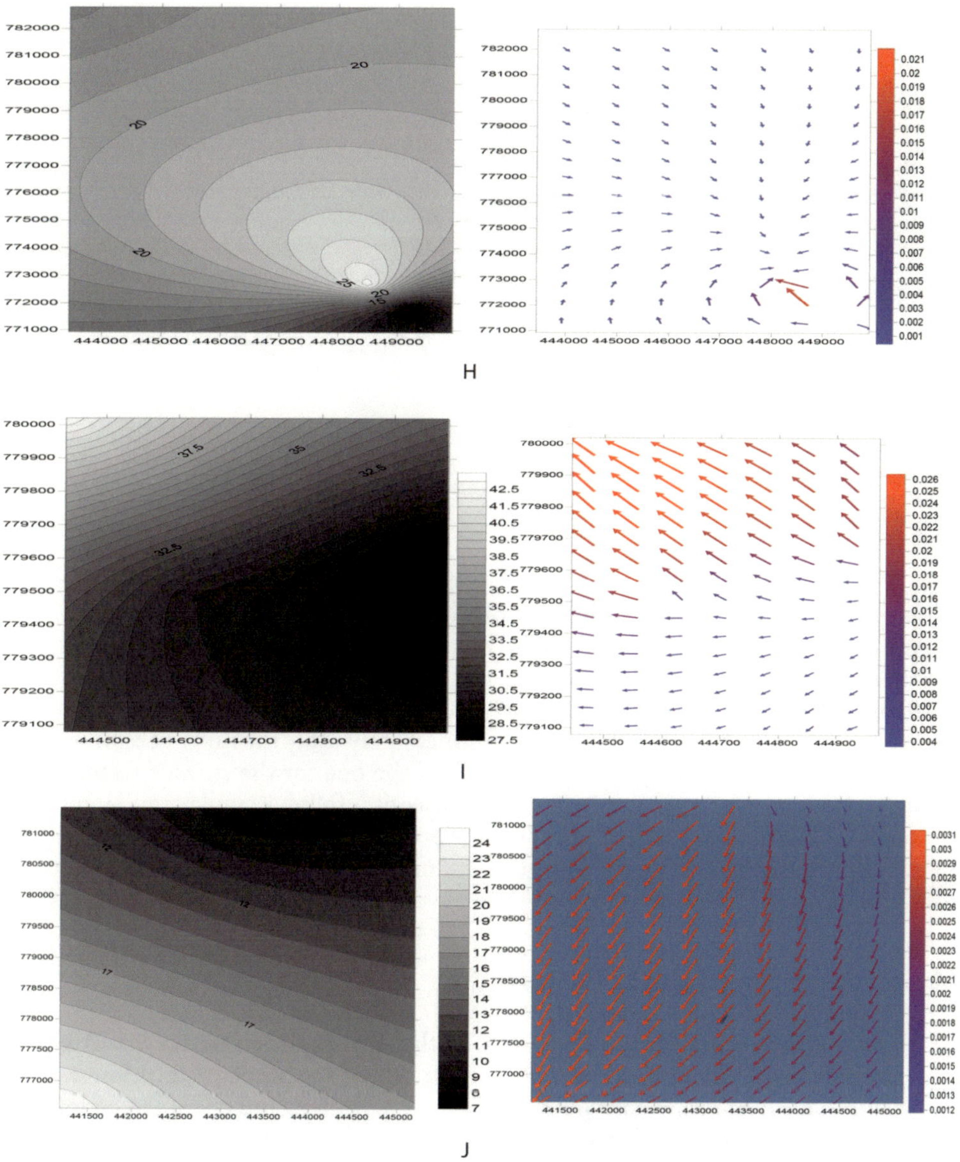

Figure 3.
(A) SWL contour map and vector map showing flow direction (2006) with color scale; (B) SWL contour map and vector map showing flow direction (2009) with color scale; (C) SWL contour map and vector map showing flow direction (2012) with color scale; (D) SWL contour map and vector map showing flow direction (2013) with color scale; (E) SWL contour map and vector map showing flow direction (2014) with color scale; (F) DWL contour map and vector map showing flow direction (2006) with color scale; (G) DWL contour map and vector map showing flow direction (2009) with color scale; (H) DWL contour map and vector map showing flow direction (2012) with color scale; (I) DWL contour map and vector map showing flow direction (2013) with color scale; and (J) DWL contour map and vector map showing flow direction (2014) with color scale.

S-W areas. For the year 2009, the map presented shows turbulence, so the flow direction has no clear trend except for the lake shore area that receives water from the nearby aquifer. For the year 2012, relatively medium magnitude water flows into the Gara Riqata area, which produces huge water discharge for the next year 2013. The two years 2013 and 2014 results clearly reveal that significant amount of

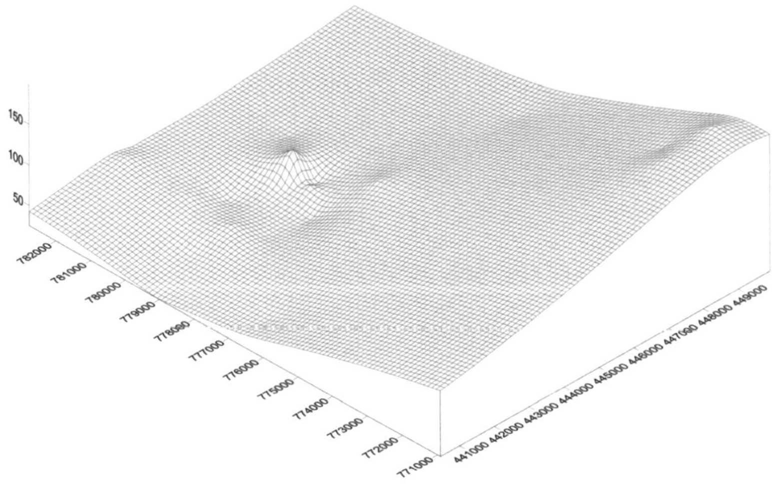

Figure 4.
Well depth wireframe map.

ground water flows from the large area of the city toward the lake shore. Thus, the ground water system feeds the lake which will in turn accelerate the lake level rise. For those months in which the pumping test of the wells conducted (i.e., March, April, and May) large amount of infiltrated water from the discharge spots join the ground water reservoir as this is the time immediately after the rainy season stops in the area. Combining these graphical results and the one for the surface water flow direction (**Figure 3A–J**), groundwater flow direction follows similar tendency of the surface water movement, that is, it follows the surface profile. This also implies the fact that the highlands are recharge areas and the lowland areas as nice spots for discharging.

2.3 Guide to drilling depth in Hawassa City

Using the wire frame and the grid map of the well depth on Surfer 8 Software, one can interpolate the required depth to be drilled at specific GPS location in the study area. This is an important guide for borehole drillers and clients to estimate the depth of the water striking formation at that specific site. To do so, the Surfer 8 Software can be used to display the map as shown in **Figure 4**.

The gridding method in Surfer 8 Software uses weighted average interpolation algorithms. This means that with all other factors being equal, the closer a point is to a grid node, the more weight it carries in determining the Z value at that grid node. The difference between gridding methods is how the weighting factors are computed and applied to data points during grid node interpolation. The coefficient of determination for this analysis is found to be $R^2 = 0.87$, which indicates the strong acceptability of the guide.

To increase the likelihood that these data are honored, one can increase the number of grid lines in the X and Y direction. This increases the chance that grid nodes coincide with data points, thereby increasing the chance that the data values are applied directly to the grid file.

The geological formation at the depth to be drilled and the other parameters determined by this and other studies shall be combined to get more detailed information. Certainly, the more the depth drilled, the better will be the safe yield.

3. Conclusion

The main aquifers tapped by the Hawassa City ground water system is unconfined and semi-confined type since no confining beds like clay are clearly identified up to 200 m depth. Weathered and fractured pumice, basalt scoriaceous rocks, fine-to-coarse-grained sand, and weathered ignimbrites are major water-bearing formations. The first water striking point is the shallowest around the lake shore (west and S-W) and as the well site goes apart to the west direction, the water striking depth is increasing and ash, fractured basalt and ash with scoriaceous basalt are dominating. The Hawassa City ground water system is of high performance and potential due to the very small drawdowns, fast recovery percentage (up to 100%), high transmissivities, and saturated thicknesses of the aquifer. The aquifer materials are highly porous and the high aquifer porosities imply aquifers of high storativity and better yield. The protective capacity of the overburden rock materials in the area is very low. Transmissivity and hydraulic conductivity values are generally high in the lake shore and central parts. Since the aquifer materials in the study area are highly permeable and relatively shallow, the groundwater has a high susceptibility of being contaminated over large area. The ground water flows from the E and S-E parts toward the central and western side of the city with a very similar profile with the surface water flow direction.

Author details

Shemsu Gulta[1] and Brook Abate[2*]

1 School of Biosystems and Environmental Engineering, Hawassa University, Hawassa, Ethiopia

2 College of Architecture and Civil Engineering, Addis Ababa Science and Technology University, Addis Ababa, Ethiopia

*Address all correspondence to: brooka12@yahoo.com

References

[1] Seifu K. Environmental isotopes and geochemistry in investigating groundwater and lake hydrology: cases from the Blue Nile basin & the Ethiopian Rift (Ethiopia) [MSc thesis]. France: University of Avignon; 2004

[2] Matthew U. Hydrogeology Lecture Notes. 2.3 ed. Bombay, India: IIT Bombay Press; 2012

[3] Kruseman GP, De Ridder NA. Analysis and Evaluation of Pumping Test Data. 2nd edition (completely revised) ed. Wageningen/Amsterdam, Netherlands: International Institute for Land Reclamation and Improvement/ TNO Institute of Applied Geoscience, Delft Institute for Earth Sciences, Free University/VU; 2000

[4] Khadri SFR. Estimation of aquifer parameters in basaltic terrain and the application of wireless sensor networks. In: Technical Report Number 2008-0917, Technical Reports and Reprints Series. Chikaldhara Region, Amravati District, Maharashtra, India. University Park, PA 16802, USA: Department of Statistics, Pennsylvania State University; 2008

[5] Cooper HH, Jacob CE. A generalized graphical method for evaluating formation constants and summarizing well field history. Transactions of the American Geophysical Union. 1946;**27**: 526-534

[6] Mishra PK, Kuhlman KL. Unconfined Aquifer Flow Theory— From Dupuit to Present. 2013. ArXiv: 1304. 3987v1. Available from: http:// arxiv.org/abs/1304.3987v1

[7] Neuman SP. Effect of partial penetration on flow in unconfined aquifers considering delayed gravity response. Water Resources Research. 1974;**10**(2):303-312

[8] Neuman SP. Analysis of pumping test data from anisotropic unconfined aquifers considering delayed gravity response. Water Resources Research. 1975;**11**(2):329-342

[9] Moench AF. Importance of the vedose zone in analyses of unconfined aquifer test. Ground Water Journal. 2004;**42**(2):223-233

[10] Moench AF, Garabedian SP, LeBlanc DR. Estimation of hydraulic parameters from an unconfined aquifer test conducted in a glacial outwash deposit, Cape Cod, Massachusetts. Professional Paper 1629. USGS, Menlo Park, CA: 2001

[11] Moench AF. Combining the Neuman and Boulton models for flow to a well in an unconfined aquifer. U.S Geological Survey. 1995;**33**(3):345

[12] Moench AF. Double-porosity models for a fissured groundwater reservoir with fracture skin. Water Resources Research. 1984;**20**(7):831-846

[13] Moench AF. Computation of type curves for flow to partially penetrating wells in water-table aquifers. Ground Water. 1993;**31**(6):966-971

[14] Moench AF. Transient flow to a large-diameter well in an aquifer with storativity semi-confining layers. Water Resources Research. 1985;**21**(8): 1121-1131

[15] Halford JK, Weight W, Schreiber R. Interpretation of transmissivity estimates from single-well pumping aquifer tests. In: Ground Water Technical Note. Vol. 44(3). USA: Montana Tech of the University of Montana, 1300 West Park, Ground Water Association; 2006

[16] Bobba G. Groundwater Development and Management of

Coastal Aquifers (including Island Aquifers) through Monitoring and Modeling Approaches. Burlington, Canada: National Water Research Institute; 2010

[17] Duffield GM. Representative Values of Hydraulic Properties. HydroSOLVE, Inc.; 2014. Available from: http://www.aqtesolv.com/versions.htm

Section 2

Water Management-Case Studies

Chapter 5

Community Management and Water Service Delivery in Africa

Joseph Okeyo Obosi

Abstract

Access to affordable and clean water has remained a challenge globally. Most states in Africa states have championed the provision of water to its citizens through state driven approaches. Despite the evidence that community water supply has contributed positively more than any other single approach to provision of water supply in Africa, it is still regarded as an informal approach. Most states in Africa still prefer other conventional approaches like Concession and Affermage in Francophone Africa and Commercialization through Management contracts in Anglophone Africa at the expense of the community water management. Either the state has not used the right approach or has neglected the community. Using evidence from Kenya, Tanzania, Malawi, Ghana and Nigeria, the paper has argued that the failure by the governments to acknowledge the disconnect between the community needs and state priorities has been responsible for the poor state of water provision in Africa.

Keywords: community management, service delivery, public private partnerships, community water, state

1. Introduction

Community water supply may acquire region or county specific definitions. However, it is generally identified by the people it serves, the purpose it serves and the manner in which it is operated. The meaning ranges from a public water system that serves at least 25 residents throughout the year and may consist of one or multiple wells or reservoirs as in Sri Lanka. Ref. [1]; an alternative to private sector participation in water supply, particularly with respect to urban areas [2] to a community-run small-scale water projects which play a crucial role in the provision of an essential service, especially in the absence of any other alternative [3]. Irrespective of the finer details of the typologies, there is a consensus on the nature and purpose of community water management. Typically, local community groups or 'community-based organizations' (CBOs) bring together people for a common purpose and are agreed on how to achieve the objective through their own management, in the pursuit of which they could be partnered in their operations by NGOs which facilitate technical and financial resources and help relax the prohibitive restrictions regarding land tenure [3]. They could also be partnered by the government or other communities for the realization of the same objective. The Community-based self-provision allows communities to form their own institutions for water delivery without formal connections or partnerships with utility operators or municipal governments. Community self-help initiatives tend to be smaller in scope than formal utility-community partnerships and often operate in small towns [4]. The rise of community water initiatives in different parts of the world gained momentum after the UNDP led Community Water Initiative (CWI) to support decentralized, demand-driven, innovative, low-cost, and community based water resource management and water supply and sanitation projects in rural areas through participatory development approach for water supply scheme planning, construction and scheme management.

From 1990s, the community water management has gained more credence as an alternative source of water supply especially in the rural areas and informal settlement areas in the urban centres. Community management of rural piped water supplies is now widely established in many countries and will become even more common in future [5]. This increasing emphasis on institutional dimensions of service delivery is also reflected in the Delhi Statement1 of 1990, which was to provide guiding principles for water supply and sanitation in the 1990s. While maintaining a focus on the use of low-cost appropriate technologies, the Statement include principles for institutional reforms, institutions of community management of services and sound financial practices [6]. In Latin America, the disadvantaged segments of the community get supply from leased regular water pipelines operated by richer businessmen on behalf of the government. In Cochabamba-Bolivia, 74% of the poorest residents lack access to municipal water service and therefore rely on communities built commonly managed wells and water systems. Although Sub-Saharan Africa is making the slowest relative and aggregate global progress with one in three people (30%) without improved drinking water access [7], most clean water is delivered via community-managed water points, either hand pumps or piped gravity-fed systems. In Bolivia the dissatisfaction of the community against privatization of water services caused serious riots that resulted into the cancelation of Multinational Water supply contract. There are scholars who have little faith in service delivery under the community management mode. In separate studies, argued that community management is less impressive than theory suggests and has serious problems have regarded the concept of community management approach as 'myth' in common pool resource management in Africa [7–10]. This is further lent credence by the fact that whether at central, regional or local, governments play dominant role in all-Africa infrastructure assessment except in water. It is only in the area of providing and maintaining water services that local communities have a leading role.

In Africa, community water supply operates mainly as an informal sector. Whereas in some countries, the supply has been a deliberate move by the government to distribute water to the disadvantaged through water communal points like in Uganda, Ethiopia and Malawi and Tanzania [11], in others especially, Kenya; community water supply has been orchestrated through self-help initiatives by local communities with no direct involvement by the government. It is prevalent in both rural and urban sectors. Once established, the community water projects seek support from donors which may include the government and its agencies; nongovernmental organizations, Church and even individuals to help them increase water access, first to the members of the organization and secondly to customers. To that extent, community water has increasingly become an alternative means to water supply to increasingly larger and economically disadvantaged segments of the society. Despite the evidence that the community water supply has contributed positively more than any other single approach to provision of water supply in Africa, it is still regarded as an informal approach.

The paper argues that community water management approach has not been pursued through its optimal level. Either the state has not used the right approach or has neglected the community. The argument is that the state ought to have used deliberate and formal approach to facilitate the management of community water supply. It is therefore the absence of a strategic approach which paved way for the community to be engaged through some semi-formal Public Private Partnership especially in Kenya.

To what extent has failure of the governments to source and distribute water to the deserving population responsible for mushrooming of community water management approach in Africa? The obtaining trend is the observation that the government is inadvertently acknowledging its inability to provide water to its citizens in the required quantity, quality and time and thereby inviting alternative suppliers/communities. Are the communities competitors or partners of the government in this endeavor? By analyzing community water management systems in Kenya, Tanzania, Malawi, Ghana and Nigeria, the paper argues that Community water management increases access to water through enhanced PPP, popular participation, and institutional governance.

1.1 Theoretical framework

The paper is based on the theory of New Public Management in the delivery of public services as propagated by Hood 1990, Kaboolian, 1998, and Page, 2005. The assumption is that governments need to disaggregate public services to their most basic units and focus on their cost management. In this context, the government shall increase access to water by recognizing and establishing community water organizations by focusing upon entrepreneurial leadership under community management, each of which will initiate their own innovations to ensure result based outputs. The community water organizations that will then be subjected to input-output control and evaluation upon performance management and audit. By doing this, more efficiency, public private partnerships and innovation shall be realized resulting into increased access to affordable water to the undeserving segment, currently unprioritized. This is informed by the argument that as currently constituted, community water management has largely been ignored. They either operate informally and independently as in Kenya, direct control under local governments like in South Africa, managed public private partnership as in Ghana or with under loosely managed and unmonitored outfits in countries where the government had initiated the projects like in Ethiopia, Malawi, Uganda and Tanzania, hence gross underperformance.

1.2 Methodological approach

The paper has used a comparative case analysis method to discuss community water governance in Kenya, Malawi, Nigeria, Ghana and Tanzania. The countries present a geographical diversity; Kenya and Tanzania in East Africa, Nigeria and Ghana in West Africa while Malawi in Southern Africa. All are part of Anglophone Africa. All except Nigeria have gone through water sector reforms and adopted mainly commercialization of water services. Whereas Kenya and Nigeria have non institutionalized community water management approach, the other three have institutionalized government driven community water management approach, with Ghana most vibrant. Tanzania and Malawi are cases of overwhelmed state projects, which are steadily giving space for private community water management interventions. Final the states form different categories of role of state in community water management. Whereas, Kenya demonstrates an Inspector/Prefect role, Tanzania and Nigeria are Mediator category while Ghana and Malawi are Benevolent States. We evaluated (1) the different sources of water for the communities, (2) The membership of community water organizations (3) the role of government in community water management and (4) The role of public private partnerships in community water service delivery. We conducted a desktop analysis of the interactions of the actors involved in community water management systems in each country ranging from source to distribution. Of particular importance was the role of the government in each of the activities and how that impacted on water access to the consumers and with what results. The fundamental question was whether or not community management improved water service delivery in the five African counties.

1.3 Results and discussion

We have discussed in this section, the dynamics of community water management in the five counties namely, Kenya, Tanzania, Malawi, Ghana and Nigeria by evaluating four processes of: sourcing and distribution of water; the role of government; the community membership; and the involvement of public private partnerships.

1.3.1 Sourcing and distribution of water

The central question was to establish how water is sourced and distributed in Africa. Different countries use different methods to facilitate water access to the community. Although the categorization of water sources usually takes the form of rural–urban community dichotomy, the sources could also be categorized as natural and artificial/conventional dichotomy. There are mainly two sources of community water supply in Africa, Artificial/Conventional and Natural. Whereas artificial sources include pipes, bore holes, wells and dams, Natural sources include rain, rivers/streams, and ponds. The uniqueness lies in the management of the distribution of the water from the sources to the consumers and the actors involved in the process. The extent to which a large population depends on natural sources is a manifestation of low level intervention of the government, thereby leaving the population to depend on nature. Whereas the natural sources are free, the water quality is unsafe and depends on climate variability. The use of conventional sources, whether fixed point or mobile vendors require investments in pipe, treatment and other means of distribution and abstraction permits for Water service providers (WSPs).

Each of the five countries, in different proportions, has both natural and artificial sources of community water supply. In Kenya, community water projects have been recognized as alternative water Service Providers (WSP) and are registered by Water Services Regulatory Board (WASREB) after meeting the conditions set by the regulator. The community water supplies are mainly through Individual bore holes, shallow wells, and water connection (kiosks) main utility companies [12]. Whereas some households have shallow wells in their yards, which neighbors are able to access free of charge, others rely on a single tap from which they sell water by the jerry can. In Dar es Salam, Tanzania, some entrepreneurs have constructed small-scale piped networks, supplying water kiosks (canteens) without a piped connection and sell water from Dar es Salaam Water and Sanitation Company (DAWASCO) even in areas that are beyond the reach of the utility's piped network. The Mobile Vendors include water tankers and trucks; pushcart and bicycle vendors who buy from a variety of water sources and resell to households. Another system is where DAWASCO supplies water to a community water public taps, managed by a Water Committee elected by the community for a three-year term [13].

Water shortages in the city of Dar es Salaam, Tanzania, had forced the community to seek for alternative ways of having clean and safe water. In other words, inadequacy in the quality and quantity of water for each of its intended purposes creates need for a communal approach as a coping response. The sources to community approach to address the unmet needs. Even in cases where the government initiated water projects for the community, the management and sustainability of the project is left at the hands of the community. More often than not, citizens cannot access water without private efforts e.g. by installing water pumps or by colluding with water utility staff, or by tapping into informal networks of neighbors, water vendors and other intermediaries. Setting up a functional community water project takes high costs, a long time, complex processes of mobilizing citizens, government authorities, NGOs, and many other stakeholders.

Citizens increasingly lose hope and trust towards the government if they have to pay a high transaction costs in terms of time, money and other resources to access official improved drinking-water sources, hence encourage either the proliferation of informal water providers as alternatives or corruption within official providers [14]. Poor communities therefore either resort to buying water from water vendors, water kiosks, and other unapproved sources which inflate their household expenditure. Whereas in settlements with piped water close by, residents walk to fetch water or pay for people to cart water to their residence, where they are far from piped water supply, residents contract small-scale suppliers who deliver water in motorized tanks. To help alleviate the deficit, the government of Ghana encourages the communities to provide services for themselves in the form of self-help projects [15].

Like in Kenya, Ghana, Tanzania and Malawi, Nigeria's main sources of community water supplies, are both natural including rain water, rivers, stream and conventional ones which include public tap, borehole, hand dug well, neighbors and Water vendors formal or informal [16]. Daily water supplies either come from the natural sources or modern/conventional supply sources. While relatively over 80% of the rural population depend on the natural sources of supplies, the urban residents are mostly served with supplies whose regularity vary depending on residential areas and other socio-economic characteristics, mostly related with ability to pay as well as the relative influence of certain individuals and groups [17]. Unlike in Kenya and Tanzania, formal water vending in Nigeria is undertaken by formal bodies, such as water utilities themselves or registered associations, or by small scale informal supplies in tankers and the water is obtained either from treated utility supplies or from registered sources. Like in Kenya and Tanzania, informal vendors in Nigeria obtain water from many different sources, protected and unprotected and deliver small quantities of water for domestic use in a variety of ways ranging from carts and cycles to containers or wheel barrows, trolleys and animal-drawn or mechanized carts and tanker trucks [18]. The government does very little to guarantee safety of the sources of community water.

Notwithstanding its intervention, Community water supply in Nigeria is still uncoordinated, mainly done by individuals to address the unmet needs by the government supplied water to the community. Unlike in Kenya, the registered associations do not necessarily supply water to the defined membership but to the market implying that those who cannot afford are still at risk of going without water. In the face of absolute neglect by the government with respect to water provision, water vendors come in as an intervention. Although source is defined as unsafe, it is unlikely that in the immediate future the government will succeed in providing adequate and safe water supply to most urban centres in Nigeria [16].

In all the countries, the government has acknowledged community water supply sources, irrespective of their safety and quality. In Ghana and Malawi, the government contributes heavily for the establishment of conventional sources. In Tanzania, the government has accommodated community identified sources. In Kenya, the choice of sources to establish is the prerogatives of each community. The fact that natural sources, most of which are unsafe, still form a large portion of the community water source, is an indication of the African governments' lethargy towards providing quality water to its citizens. The use of different sources of community water is therefore a coping mechanism to mitigate the failures of the governments to provide adequate water.

1.3.2 The role of the government

Different governments play different roles in community water supply in Africa. In assessing the role of government, we targeted not only the regulatory roles but also mobilization of the community, financing the development and maintenance of the water sources.

The government of Kenya performs the following responsibilities in as far as community water projects are concerned: issue of permits for sinking of bore holes; registration of water service providers (WSPs); payment of water abstraction fee and regulation of water tariffs and quality. The Water and Sanitation Regulatory Board (WASREB) oversees, on behalf of the Ministry of Water and Irrigation, the implementation of the National Water Services Strategy and Pro-Poor Implementation Plan, which specifically focuses on expanding services to under-served low-income areas in Kenya.

The government of Tanzania initially provided water directly to communities by facilitating sinking of boreholes to residents of a locality before the strategy was overwhelmed and the government warmed up to private initiatives. The Water Supply and Sanitation Act 2009 provided room for the formal establishment of Community-owned Water Supply Organizations in various forms operating around the aegis of: Dar es Salaam City Council (DCC); the civil societies; political party organizations and private individuals as well as youth and women groups; and the donor community which provided facilities to fringe settlements. The organizations individually operated water-kiosks or boreholes in informal or peri-urban settlements, initially constructed by DAWASCO or by NGOs such as Water Aid and PLAN International. The flexibility is intended to allow them to build on trust and integrity already developed through existing social networks. According to the Water Policy, grassroot institutions linked to local government including the Village/Mtaa Water Committees (VWCs), are responsible for the management of water supply schemes in their localities. Recognition of these structures by the government through DCC gave the Mtaa some power of implementing community action plans prepared by the various committees. The Mtaa leadership had been playing a vital role in mobilizing resources and organizing community participation and private involvement in water provision to the fringe areas [19].

Similarly, the government of Ghana established Community Water and Sanitation Agency (CWSA) in 1998 by an act of Parliament (CWSA Act 564) to provide the institutional base for the implementation of the national community water and sanitation programme [20]. In conjunction with District Assemblies, the CWSA developed procedures which Communities applying for water and sanitation facilities should follow including composition of membership and proportion of women in the membership and bank account before the grant could be provided. This was to facilitate access to water for those without direct connection to the state owned Ghana Water Company Limited (GWCL) and continue to rely on informal services or secondary and tertiary sources due to reasons including lack of land titles or non-affordability of the upfront connection fees [21]. The strategy was to involve the community in planning and management of their water supply systems and water resources through their elected Water and Sanitation (WATSAN) Committees, Local Water Boards, Water and Sanitation Development Boards, Unit Committees and Area Councils [22].

Although not necessarily to the same extent as Ghana, the government of Malawi also facilitated community water supply through Water service boards and establishment of Water Users Association (WUA). The government owned water management Boards own the main water infrastructure (treatment plants, etc.), and communities own public standpipes or water kiosks that are managed by their respective WUAs.

The National Rural Water Supply and Sanitation Policy, 2000 enhanced the role of government of Nigeria at the centre of community water management by emphasizing rural water and sanitation through community participation. In most cases, Water Boards or Water Corporations are used at the state level for urban water services while rural water supply and sanitation (RWSS) is used for rural water supply and sanitation. All the 774 local government authorities are further involved in the provision and management of rural water supply and sanitation within their respective domains, mostly through various community organization including water and sanitation committees (WASCOMS).

In all the countries except Nigeria, community water supply had institutional domains by either being domiciled in Ministries of water and Health or equivalents. In Nigeria, different Ministries and Agencies assume relevance and arrogate water and sanitation responsibilities for their respective Ministries without clear mechanism of coordination. At the state and local government levels, there are further fragmentation and division of authorities to the extent that what emerge are inter-agency competition both between agencies of each state and between agencies of States and the Federal Government. This consequently leads to parallel drinking water projects in some areas and communities as well as duplication of responsibilities. Allocation of water and sanitation projects is often politicized to favor communities with influential public officials, bureaucrats or politicians [16]. Whereas there is direct intervention by the government in community water supply in Ghana, Malawi, Tanzania and Nigeria, in Kenya, there is indirect intervention with a lot of self-initiatives. In Nigeria, government has a selective intervention and makes little attempt at mobilization of community participation.

1.3.3 Membership to community water projects

The membership to community water projects depends on the type of supply. For individual and private water sources, there is no standing membership save for the payment to the owners. Most of community water projects in Kenya are communally owned by individuals who came up together and formed an association to source for funds to supply water to its members, usually those in the neighborhood. With time they expand to access other non-members through payment for water. The access could be either through water stand points or connection to individual households. The associations could equally draw water from main water utility company and set up a bulk meter outlet from where they could establish other connections. In a way, they serve as distributing agents at a profit on behalf of the main water companies (Obosi).

The membership to community water projects in Tanzania is tied to area of residence, or village commonly known as "Mtaa", an extension of local government. Each village committee elects leaders who mobilize the community to run the government-sponsored water project. Water is managed by water committees and overseen by the political leaders such as street chairpersons. In Dar es Salaam, the performance of community water projects in ensuring reliable access of water to citizens greatly depend on the performance of local political leaders [13].

Like in Tanzania, membership to community water projects in Ghana are also area specific and is indirectly driven by the state which has stipulated conditions including quota for women and initial 5% of the cost, for support to establish community water supply through the District Assemblies and CWSA. Community projects formed this way receive government support to the tune of 95%. Most of them (56.3%) were jointly initiated by the government (represented by the District Assemblies and the CWSA) and the community. The community water projects, in reality, are largely donor-sponsored projects, which were implemented by the government through the CWSA [14].

Membership of community water projects in Malawi is tied to proximity and is state driven through Water Users Associations (WUA). Each community has one WUA that is expected to represent the interests of all water users. The utility provides technical expertise to WUAs through trained plumbers and other certified workers. WUAs collect revenue from water sales and pay the utility on a monthly basis. The Water Board (WB) benefits from this organized, streamlined revenue collection system and can operate more efficiently, not having to supervise and pay for employee (vendor) salaries. Each WUA typically comprises an appointed board of trustees that is the final decision-making body and disciplinary arm; an executive committee that is voted into office to oversee the day-to-day running of the association; a secretariat with employees headed by an administrator; inspectors who audit water meter readings and report faults to the WUA office; and water vendors who sell water at the kiosks. Community member customers contribute financially to the WUAs by purchasing water and participate in the election of executives.

The fact that all the four countries have relied on government for community water supply yet they still have challenges is an indication that reliance on the government by residents may not guarantee them unlimited access to water supply. Like in Kenya, there is need to engage in self- help water supply projects by pooling resources together to either sink boreholes or dig wells for their common use. This should be coupled with community involvement in the water management process, which is existent in all the countries except Nigeria. This will help residents to have a sense of responsibility and thus curb the occurrence of vandalism of water equipment [17].

1.3.4 Involvement of public private partnership

Public private partnership is an arrangement in which the private sector in form of individuals, corporates or community get into a co-production with the state through shared responsibilities. In water supply the arrangement have included Public Enterprise, Joint Ventures, Affermage, Built Operate Transfer, Built Operate Transfer, Concession, Lease, Management Contracts and Private Ventures [12]. The timing and extent is a prerogative of the partners. Most of community water projects in Kenya operate through public private partnerships. In addition to engaging government for regulatory services, they also apply for government funds as self-help projects either through constituency Development Funds (CDF) or through Water services Trust Fund (WSTF) mainly to improve their infrastructural development for uptake, supply and storage. These are usually one off assistance and no compelling continued partnership. However for the funding from the WSTF, the condition for qualifying for the grant is access to the rest of the community. In the urban areas, some utility companies, like Kisumu Water and Sanitation Company (KIWASCO) in Kisumu, have as part of its approach to providing services in low-income areas of Kisumu implemented a delegated management model since 2004, in which it partners with small-scale private operators which are formed from within the community that is to be served [23].

Under this model "the utility sells bulk water to an agent who has been contracted to operate and manage part of the system". These agents who operate and manage the last part of the service delivery system are known as 'Master Operators' (MOs) [6].

The first appearance of PPP in water sector in Tanzania was the setting up of Water kiosks by DAWASCO as a means to supply water to communities that do not have in-house water connection. The kiosks are run by private agents who in return pay rent to DAWASCO [24]. The other aspect of Partnerships involved drilling of boreholes in which both public and private water drilling companies mediate citizens' access to water. The partnership usually brings together public services and NGOs to provide water [25]. The strongest move towards actualization of PPP in Tanzania was setting up of a Community Liaison Unit by DAWASA in 2003 to help community-managed suppliers [26]. Various resources such as expertise and experience, ideas, ability to organize, materials, labour and finance were mobilized and used to improve potable water in informal settlements through participation of various grassroots and external actors were involved. The grassroots actors include the water users, i.e. individual households, vendors and the Mtaa leaders. External actors came in as advisors, financiers and contractors in the construction of the wells and include DCC, DAWASA, the Water Resource's Institute (WRI) and the UNDP and Lions Club. While DCC has been acting as a facilitator, DAWASA staff have been providing free professional support in the form of, for example, amount of chemical materials for treating water, to the WMCs and to individual private owners. The community, private individuals (vendors), government institutions, training institution, NGOs, e.g. Lions Club, local informal and formal businesses and the United Nations Development Programme (UNDP). The community had contributed funds through the initiation and co-ordination by the then village government (known also as CCM leadership) of the area - financial assistance from the Lions Club of Dar es Salaam and technical assistance from the Water Resource Institute also of Dar es Salaam.

In 2006, the government of Malawi, in an attempt to address chronic water scarcity in urban informal settlements, promoted community-public partnerships (CPPs), a form of service co-production in which state water utilities work together with community-elected water user associations (WUAs) [27]. The CPPs involved community elected representatives and state water-delivery agencies over an indefinite period: communities primarily oversee the management water services and revenue collection, while the utility manages infrastructure, delivers water to community pipes, and provides technical assistance [27]. Nongovernmental organizations, Water Aid and the Centre for Community Organization and Development (CCODE), and the Lilongwe Water Board (LWB), a public water utility, mobilized community leaders to form Water User Associations (WUAs). The partnerships are intended to enhance water supply in underserved urban settlements and create opportunities for communities to participate actively in water service delivery. Community-elected WUAs manage revenue from water sales, oversee community water points (kiosks), organize community elections to appoint representatives, and report community complaints about service delivery to the LWB. LWB, in turn, supplies water to WUA communities, provides technical assistance, and manages Lilongwe's main water infrastructure.

Private sector participation in the Rural Water and Sanitation (RWSS) sub-sector in Nigeria has been in the form of consultants, suppliers, manufacturers, artisans. Non-Governmental Organizations (NGOs) are even becoming equally relevant in the RWSS through collaboration with relevant authorities, communities and donor organizations including: UNICEF assisted State Water and Sanitation Projects (1981–2010); Japanese International Cooperation Agency's (JICA) rural water supply projects (1992–1994); United Nations Development Project (UNDP)-Rural Water supply (1988–1993); European Union (EU) water and sanitation programme (2002–2009); Department for International Development's (DFID) water and sanitation pilot project (2002–2008); Water Aid's rural water supply and sanitation programme (1996–2010); United State Agency for International Development; World Health Organization and World Bank [28]. Involvements of these bodies have been restricted to financing, infrastructural provisions in urban, rural areas and public spaces.

1.4 Analysis

The discussion of the results shows that different countries in Africa have used community management differently for various reasons and in varying degree of success. In all instances, community water management has been used as an informal approach especially where the mainstream approaches have not been able to access. Even in countries like Ethiopia and Malawi where the government established the community water supply, there is still little faith in its management [5].

The exponential of growth of different sources of water though at face value, creates an impression of improved water supply, the dependence on natural sources of water by a large population depicts a gloomy picture of access to safe, quality and reliable water. It means that a large population is still vulnerable to waterborne diseases arising from unprotected sources of water, hence not only giving the government a temporary reprieve but also a false hope of less pressure from citizens for water. The trend is worrying since even countries which began with benevolent approach of supplying water to the communities like Ghana, Tanzania and Malawi have ended up being overwhelmed by demand and had to change strategy to accommodate more initiatives from the community. This means that it is not just about the state's direct involvement but taking the lead in promoting constructive engagement based on each community's need, without treating the communities as homogeneous entities. The fact that the communities still rely on natural sources of water is a further manifestation of the projects being initiated in desperation and as a coping mechanism. It does not mean that they are capable of producing reliable quality water at the expense of the government. The arguments of some opponents of community water supply that emphasizes government's attempt to run away from its role and that the approach is not sustainable, hence suffice. However, opponents of community management argue that the model is neither cost effective nor sustainable hence does not work well for communities due to various reasons including: non functionality of many such water points do not work by roughly one-third across the continent [29]. In Tanzania, one-quarter of new water points become non-functional within 2 years of installation [14]. For related reasons some scholars argue that community management is the least preferred management option for water users [10]. Other studies though appreciating the role of community water supply, established that the government's preferred choices in the management especially of maintenance is at times at variance with that of the community, hence less gain [10]. In Malawi, the technical and financial performance under community management is weak and therefore the community management has worked more for the state and donors as a means of offloading public service delivery responsibility than it is for the community and therefore cannot deliver the desired results (Elly [8, 9]).

In terms of roles of government in the community water supply in Africa, three broad categories are identifiable; The Prefect/Inspector; The Mediator and the Benevolent. In the prefect category, the government seldom mobilizes the community. Even where it does, its sole preoccupation is whether the community is following the prescribed procedures. Although it might not bother so much even if an initiative sprouts from a community, however, that initiative must comply with the law for it to be permitted to work, failure to which it is branded informal and its water unsafe. Due to its inability to comply with its obligation of providing an alternative, the government is embarrassed to either stop the operations or help in the system improvement. It may however put some stringent pre-conditions for community projects to qualify for the government support. The resultant scenario is the mushrooming of so many unregulated sources of water supply including natural sources, illegal tapping of government utility water companies. Kenya leads in this kind of category. The Mediator category, both Passive and Active types, provides an institutional framework to facilitate community water framework. The passive mediator, may negotiate general support with donors and development partners but does not enforce the support to the individual community water providers. It is upon any individual entrepreneur to grab the opportunity and supply water to the deserving community. This results in uncoordinated approach usually resulting in exploitation of the underprivileged community and differential distribution of water by privileges and status like the case is in Nigeria.

The Active mediator type government accepts responsibility to provide water and after being overwhelmed, it not only creates institutional framework for support also but encourages donor partnership with the community. The communities are not compelled to embrace donor support through mobilization and creating space for community-donor engagement in the implementation process through an administrative forum. This is a case that obtains in Tanzania, where the community leadership and the donor meet under the state facilitated Village/Mtaa Water Committees. The arrangement does not interfere with other forms of community water provision, either through vendors or private fixed water points. They operate side by side. Neither does the government restrict individual and private initiatives to provide water through other informal means.

The third category, the Benevolent State presents a situation in which the government deliberately creates structure for community water supply. It is the responsibility of the state to design and provide water access points for the community and is coordinated by the District Assembly/Local Administration. The community is organized around known structures and area of a particular number of households, who are mobilized to form Water User Associations for the management of the centralized community water supply. The government determines the membership. This category obtains under community water management approach in Malawi and Ghana and earlier own Tanzania before it slipped into the Mediator category. In Malawi and Ethiopia, each government not only designed but also constructed water points before inviting communal involvement. Ghana formed National Community Water and Sanitation Programme (NCWSP) to facilitate the provision of basic water and sanitation services to communities through Community Ownership and Management [30]. Even though the government of Kenya has strengthened the legal basis and capacity of community-based service providers, they are still regarded as informal or small scale water service providers. Whereas in some counties, the supply has been a deliberate move by the government to distribute water to the disadvantaged through water communal points like in Uganda, Ethiopia and Malawi, in Kenya, community water supply has been orchestrated through self-help initiatives by local communities with no direct role by the government.

The involvement of Public Private Partnership in the community water management in Africa is very prominent, the extent to which differs from state to state. Irrespective of the success, its emergence was no doubt, occasioned by the realization that neither the government nor the private sector alone could provide quality and reliable water in good quantity and time to the community in Africa. The difference in extent of involvement of PPP is related to the category of role of governments. In the Benevolent category, the state champions the search for strategic partners to establish the infrastructure and mobilize the community to manage the community water supply. This challenge has generally been hampered by sustainability problems after the partners have left especially in Malawi. It is however less in Ghana due to stronger institutional governance support. In the Passive mediator category, there is limited activity in partnerships since only those championed by the state are active and sustainability challenges arise shortly thereafter. In the active mediator level there, are more PPPs both initiated by the state and by individuals. There is a flurry of PPPs in the Inspector/Prefect Category where individuals, private corporates and the government all participate asymmetrically. There is no predetermined or prescribed way of partnerships. This is consistent with other scholars observation. In Malawi and Ethiopia, the respective governments deliberately both singly and in support of International NGOs, established community public water standpipes to provide access to rural population to water. In Kenya community management was even stronger and started through self-help initiatives and for members first. To date community water supply contributes up to 60% of total water access in Kenya [12], 40% of access in Dar es Salaam [31]. Of the 8 million Kenyans who have access to improved water in rural areas, 30% are served by community-managed water supply schemes most of which were developed by self-help groups. These self-help schemes differ from those in Ethiopia or Malawi in two important aspects. First, they were designed to provide water mainly to the members of the self-help groups, not equitably to everybody living in the service areas. Secondly, they supply water mostly through household connections, not public tap stands. Only 26.8% of the population have access to a basic minimum level of service in Ethiopia, while 64.2% have access in Kenya and 84.5% have access in South Africa (WSP, 2003). Whereas Central government is highest water sector provider at 51%, followed by Local authorities at 27% and Non-Governmental organization including CBOs and PSP at 21% in Kenya, In Ethiopia Private Sector Participation through CBOs is at 54% followed by local authorities at and no central government direct participation.

Like in Ghana, the governments of Ethiopia and Malawi worked in partnership with the communities and, the former providing technical standards and supervision. The government took the lead in implementing projects, and then in the 1990s Water Aid, the international NGO, began giving financial and professional help to the government schemes. The government engineers designed the schemes in accordance with technical standards and the wishes of the communities served. Ghana has further institutionalized Public Private Partnership in which involving contracted Private Operators and Public Operators under Community Ownership and Management approach under the supervision of local authorities through District Assemblies. However, in Ethiopia, the Ethiopia Social Rehabilitation and Development Fund (ESRDF) provide grant funding through the national budget and the communities cover 10% of capital costs and all operating costs. As in Ethiopia, the projects in Malawi were designed to serve the entire population in the supply area, but only through public tap stands.

Table 1 shows that the growth of community water projects in inversely related to the role of state. When there are more projects, the role of state tends towards that of an inspector. When the reverse is the case, the role of the state tends to be more of benevolent. Even the states whose role began as benevolent, eventually need to tend towards mediator and later Inspector to ensure quality and safety.

1.5 Conclusion

The community water management supply has emerged as a core intervention strategy in Africa to fill in the space the state has either abandoned or unable to occupy. It has emerged and matured in various ways in different countries ranging

Role of state in community water	**Kenya**	**Tanzania**	**Nigeria**	**Malawi**	**Ghana**
	Inspector	**Active mediator**	**Passive mediator**	**Benevolent**	**Benevolent**
Water sources	Community-operated kiosks; shallow wells and boreholes; natural sources i.e. rain water, rivers, lake, ponds; and vendors	Community-operated kiosks; shallow wells and rivers; vendors and mosques	Community-operated kiosks; shallow wells and boreholes; natural sources- rain water, rivers, lake, ponds, and vendors	community-operated kiosks; natural sources; shallow wells and boreholes	Community-operated kiosks; shallow wells and boreholes
Role of state	Regulation, Ltd. infrastructure development; collecting fees and registration of WSPs	Mobilization of community and partners; infrastructure development	Mobilization for infrastructure development	Mobilization of community and partners; infrastructure development	Mobilization of community and partners; infrastructure development
Membership	Local community groups; community-based organizations' (CBOs); institutions; welfare associations/ organized groups	People living a given radius; individuals and NGOs	People living a given radius; individuals	People living a given radius-initiative of the community	People living a given radius-initiative of the community
PPP	Donors, community, state; WSP	Donors; WSPs; water utilities; NGOs; community liaison unit, political party leadership, local authority	Consultants; non-governmental organizations; local government authorities; local community water committees; donors	Donors, nongovernmental organizations; center for community organization and development (CCODE); water board; WSP; local community/ WUAs	Donor, community, state

Direction of flow community water and state involvement. ←

Table 1.
A framework of the involvement of state in community water management in Africa.

from state supported in Ghana, Tanzania, and Malawi to amorphous in Nigeria to self-supported initiatives in Kenya. Even where the state has supported the initiatives, the state has been overwhelmed more informal water supplies still emerge as in Nigeria and Tanzania and ended opening up to more public private partnerships like in Ghana. The higher the number of community water projects are, the more likely the state will assume the Inspector role. There are fewer community water projects in benevolent states. In all three categories of roles, ranging from

Benevolent, Mediated and Inspector/Prefect, it is a clear that the state in Africa cannot run away from facilitating community water supply. Neither can it wish it away as long as citizens continue to go without water. In fact, the community water and the state are strange bedfellows in the water supply. At times they act as partners, and at times as competitors especially in Inspector states. Since neither can do without the other, public private partnerships is the best option. The countries that have demonstrated high level of PPP have equally been able to relatively, show more improvement in the water supply. In order to address inequality, quality and exploitation the state still has complement the mobilization of resources by the private sector.

Author details

Joseph Okeyo Obosi
Department of Political Science and Public Administration, University of Nairobi, Kenya

*Address all correspondence to: jobosi@uonbi.ac.ke; jokeyoobosi@gmail.com

References

[1] Mimrose D, Gunawardena E, Nayakakorala H. Assessment of sustainability of community water supply projects in Kandy district. Tropical Agricultural Research. 2012;**23**:51-60. DOI: 10.4038/tar.v23i1.4631

[2] Bakker KJ. The ambiguity of community: Debating alternatives to private-sector provision of urban water supply. Water Alternatives. 2008;**1**(2):236-252

[3] Dagdeviren H, Robertson SA. Access to water in the slums of sub-Saharan Africa. Development and Policy Review. 2011;**29**(4):485-505. DOI: 10.1111/j.1467-7679.2011.00543.x

[4] Adams EA, Sambu D, Smiley SL. Urban water supply in sub-Saharan Africa: Historical and emerging policies and institutional arrangements. International Journal of Water Resources Development. 2019;**35**(2):240-263. DOI: 10.1080/07900627.2017.1423282

[5] Water Sanitation Programme. In: Jon L, editor. Nairobi: World Bank; August 2002. Retrieved from: www.wsp.org

[6] Schwartz K, Tutusaus M, Savelli E. Water for the urban poor: Balancing financial and social objectives through service differentiation in the Kenyan water sector. Utilities Policy. 2017;**48**:22-31. DOI: 10.1016/j.jup.2017.08.001

[7] Wunsch JS. Decentralization, local governance and "recentralization" in Africa. Public Administration and Development. 2001;**21**(4):277-288. DOI: 10.1002/pad.185

[8] Chowns E. Is community management an efficient and effective model of public service delivery? Lessons from the rural water supply sector in Malawi. Public Administration and Development. 2015;**35**:263-276. DOI: 10.1002/pad.1737

[9] Chowns E. Water Point Sustainability and the Unintended Impacts of Community Management in Malawi. Water, Sanitation and Hygiene Services Beyond 2015: Improving Access and Sustainability; 2015

[10] Hope R. Is community water management the community's choice? Implications for water and development policy in Africa. Water Policy. 2015;**17**(4):664-678. DOI: 10.2166/wp.2014.170

[11] Bayliss K, Tukai R. United Nations Development Programme. Services and supply chains: The role of the domestic private sector in water service delivery in Tanzania. 2011. pp. 1-40. Available from: http://www.undp.org/content/dam/undp/library/PovertyReduction/Inclusive development/Tanzania-Water.pdf

[12] Obosi JO. The Public Service Delivery Challenge: A Public Private Partnership in Water Service Provision in Kenya. 1st ed. Saarbrucken: Schaolars' Press; 2015. Available from: www.omniscriptum.com

[13] Dakyaga F, Kyessi AG, Msami JM. Water access today and tomorrow: Domestic water sustainability under informal water supply markets in Dar es Salaam, Tanzania. Journal of Sustainable Development. 2018;**11**(6):120. DOI: 10.5539/jsd.v11n6p120

[14] theodory T, Malipula M. Perceived impact of private sector involvement in water supply on the urban poor in Dar es Salaam. Tanzania Journal of Development Studies. 2013;**13**(1-2):96-108

[15] Crook R, Ayee J. Urban service partnerships, 'street-level bureaucrats' and environmental sanitation in Kumasi and Accra, Ghana: Coping with organisational change in the public bureaucracy. Development and Policy Review. 2006;**24**(1):51-73. DOI: 10.1111/j.1467-7679.2006.00313.x

[16] Akpabio EM. Working paper series. Accounting and Finance. 2009;**24**(1): 75-75. DOI: 10.1111/j.1467-629x.1984.tb00054.x

[17] Ahile SI, Udoumoh EF, Adzande P. Residents coping strategies with water scarcity in Makurdi town, Nigeria. Mediterranean Journal of Social Sciences. 2015;**6**(4S2):100-108. DOI: 10.5901/mjss.2015.v6n4s2p100

[18] Olajuyigbe AE, Rotowa OO, Adewumi IJ. Water vending in Nigeria—A case study of Festac town, Lagos, Nigeria. Mediterranean Journal of Social Sciences. 2012;**3**(1):229-239. DOI: 10.5901/mjss.2012.03.01.229

[19] Kyessi AG. Community-based urban water management in fringe neighbourhoods: The case of Dar es Salaam, Tanzania. Habitat International. 2005;**29**(1):1-25. DOI: 10.1016/S0197-3975(03)00059-6

[20] Essegbey G, Godfred F, Entsua-mensah RM, Engmann C. Assessment of Community Water and Sanitation in Ghana. Policy Studies. 2007. Available from: http://www.atpsnet.org

[21] Monney I, Buamah R, Odai S, Awuah E, Nyenje PM. Evaluating access to potable water and basic sanitation in Ghana's largest urban slum community: Old Fadama, Accra. Journal of Environment and Earth Science. 2013;**3**(11):72-80. Available from: http://iiste.org/Journals/index.php/JEES/article/view/8254

[22] Beck T, Rodina L, Luker E, Harris L. Institutional and Policy Mapping of the Water sector in South Africa; 2016. pp. 1-12

[23] Obosi JO. Nature and scope of public private partnerships in the water sector in Kenya. Open Journal of Political Science. 2018;**8**:12-34. DOI: 10.4236/ojps.2018.81002

[24] Nganyanyuka K, Martinez J, Lungo J, Georgiadou Y. If citizens protest, do water providers listen? Water woes in a Tanzanian town. Environment and Urbanization. 2018;**30**:613-630. DOI: 10.1177/0956247817744942

[25] Kjellén M. From Public Pipes to Private Hands: Water Access and Distribution in Dar es Salaam, Tanzania. 1st ed. Stockholm, Solna, Sweden: DocuSys AB; 2006

[26] Allen A, Hofmann P, Mukherjee J, Walnycki A. Water trajectories through non-networked infrastructure: Insights from peri-urban Dar es Salaam, Cochabamba and Kolkata. Urban Research and Practice. 2017;**10**:22-42. DOI: 10.1080/17535069.2016.1197306

[27] Adams EA, Boateng GO. Are urban informal communities capable of co-production? The influence of community–public partnerships on water access in Lilongwe, Malawi. Environment and Urbanization. 2018;**30**(2):461-480. DOI: 10.1177/0956247818792059

[28] Ahmad MT, Daura HA. Domestic water consumption under unimproved water supply in semi-arid region of northern Nigeria. Sustainable Water Resources Management. 2019;**5**(2):415-424. DOI: 10.1007/s40899-017-0177-5

[29] RWSN. Myths of the rural water supply sector. Rural Water Supply Network Perspectives. 2009;**4**:1-7

[30] Nyarko KB, Oduro-Kwarteng S, Owusu-Antwi P. Local authorities, community and private operators partnerships in small towns water service delivery in Ghana. Physics and Chemistry of the Earth. 2011;**36** (14-15):1078-1084. DOI: 10.1016/j.pce.2011.08.007

[31] Mcgranahan G, Kjellén M. Informal water vendors and the urban poor. Human Settlementss Discussion Paper. 2006;**3**:1-24

Chapter 6

Effects of Climate Change on Water Resources, Indices, and Related Activities in Colombia

Nathaly Güiza-Villa, Carlos Gay-García and Jesús Efren Ospina-Noreña

Abstract

In Colombia, a country with great climatic diversity, the water balance is affected in one way or another by climate change depending on the region. Thus, there may be increases and decreases in precipitation and, in all cases, a huge increase in temperature. This document presents some studies carried out in different areas of the country regarding the effects of climate change on water resources, including its influence on hydroelectric power generation, some changes in the water balance in arid areas, and the opportunity to ensemble climate change scenarios. Likewise, it outlines a possible future water supply-demand relationship, where supply is associated with a change in the water balance and demand with some crops, activities, and sectors that need water to survive. This allows to estimate some future status indices to see the overall picture of climate change in connection with the country's water resources.

Keywords: climate change, water resources, indices, water supply, water demand

1. Introduction

Colombia is a country located in South America within the Intertropical Convergence Zone (ITCZ), which is associated with east trade winds. And, in addition to its varied orography, it boasts a great diversity of climates and rainy and dry seasons, once or twice a year, depending on the area of interest [1]. The same happens to the influence of climate change in the country, that is, precipitation varies in different ways. Moreover, during Colombia's Third National Communication on Climate Change before the United Nations Framework Convention on Climate Change (UNFCCC), the Institute of Hydrology, Meteorology and Environmental Studies (IDEAM) said that there would be a reduction in precipitation between 10 and 30% for 27% of the country, specifically in the northern and southern areas of the Colombian territory. Conversely, the same variable is expected to increase in 14% of the territory, in the central and western areas of the country [2].

However, IDEAM studies cover a very large region of the country, which makes them a poor reference for making local or scale decisions regarding the watershed. Therefore, different studies have been carried out on that scale for different regions of the country, with different climatic and geographical conditions, some of which will be detailed throughout this chapter.

2. Models and scenarios

The Intergovernmental Panel on Climate Change (IPCC) has demonstrated—with 95% of certainty—that human beings and their activities are the leading cause of global warming, which has become evident for more than five decades due to the increase in the concentrations of greenhouse gases (GHG). This will produce future changes in extreme weather and climate events throughout the planet. Therefore, the temperature and frequency of extreme events associated with precipitation, both floods and droughts, are very likely to increase progressively [3].

In consequence, a range of future climate scenarios was created in order to establish the potential influence on climate change on the planet in the short, medium, and long term. The SRES scenarios (A1, A2, B1, and B2) [4] used for the fourth assessment report (AR4) depend on a combination of future prospects for economic and technological development, and population growth. And the RCP scenarios or Representative Concentration Pathways (2.6, 4.5, 6.0, and 8.5), used for the fifth assessment report (AR5), are associated with greenhouse gas emissions measured as carbon dioxide [3]. The scenarios mentioned are briefly described in **Table 1**.

The indicated table establishes the relationship between the SRES and RCP scenarios with a possible equivalence between them, which can be used as a reference to observe the possible future changes in each of the studies mentioned in this

Scenario	Description
SRES A1	It is a world with rapid economic growth, where the population grows to some extent in the middle of the century and with a rapid spread of efficient new technologies [5]
SRES A2	It is a self-sufficient world, with continuous population growth and economic development associated with technological changes [5]
SRES B1	It is a world with a rapid change in economic structures, where the population grows to some extent in the middle of the century and with the introduction of efficient resources and technologies [5]
SRES B2	It is a world with a local economic, social, and environmental emphasis, with sustainable development, progressive population growth, and economic development oriented toward environmental protection [5]
RCP 2.6	An increase in CO_2Eq emissions is expected for the year 2100 until reaching a concentration of 490 ppm, with a peak prior to that year and a subsequent decrease, which would increase temperature from 0.3 to 1.7°C [2]
RCP 4.5	An increase in CO_2Eq emissions is expected for the year 2100 until reaching a concentration of 650 ppm, with a subsequent stabilization, which would increase temperature from 1.1 to 2.6°C. It is equivalent to SRES B1 [2]
RCP 6.0	An increase in CO_2Eq emissions is expected for the year 2100 until reaching a concentration of 850 ppm, with a subsequent stabilization, which would increase temperature from 1.4 to 3.1°C. It is equivalent to SRES B2 [2]
RCP 8.5	An increase in CO_2Eq emissions is expected for the year 2100 until reaching a concentration of 1370 ppm, with subsequent growth, which would increase temperature from 2.6 to 4.8°C [2]

Table 1.
Description of climate change scenarios.

DOI: http://dx.doi.org/10.5772/intechopen.90652

Models	Scenarios	Period	Location	Study
HadGem2-ES, GDFL-CM3	RCP 4.5 and RCP 8.5	2050-2070	Uribia, La Guajira	[6]
CCSRNIES-A21, CSIROMK2B-A21, CGCM2-A21, CGCM2-A22, CGCM2-A23, HadCM3-A21, HadCM3-A22, HadCM3-A23, HadCM3-A2-SDSM.	SRES A2 and B2	2010-2100	Sinú-Caribe Basin	[7–10]
BCCCM1-1, CCSM4, GISS-E2-R, HadGEM2-AO, HadGEM2-ES, IPSL-CM5A-LR, MIROC-ESM-CHEM, MIROC-ESM, MIROC5, MRI-CGCM3, NorESM1-M.	RCP 2.6, 4.5, 6.0 and 8.5	2050-2070	Nilo, Cundinamarca	[11, 12]
IPSL-CM5A-LR, IPSL-CM5A-MR, MIROC 5	RCP 4.5 and RCP 8.5	2020-2050	Coello River Basin, Tolima	[13]

Table 2.
Models and scenarios according to the reviewed study.

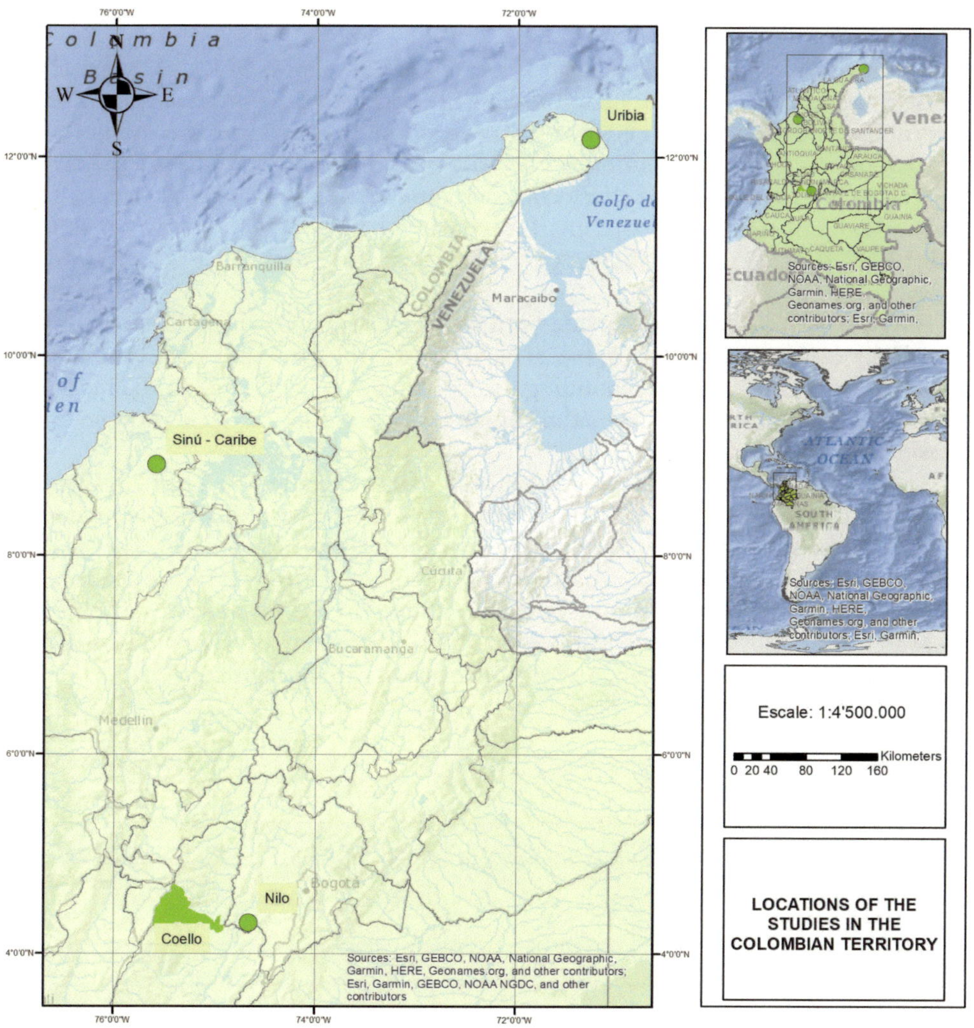

Figure 1.
Locations of the studies in the Colombian territory.

document. Given its broad temporal spectrum, **Table 2** shows the different scenarios over different periods of time.

The studies are located in different areas of the country, with different characteristics. The map illustrated in **Figure 1** shows their exact location within Colombia.

3. Effects of climate change on water resources, indices, and related activities in Colombia

This section introduces the studies carried out in different areas of the country, using different methodologies, which are explained in depth in each of the investigations. Additionally, it sets out the possible effects of climate change on the water resources in each location.

3.1 Effects of climate change on indices and climate zonification

3.1.1 Uribia, La Guajira

This study was carried out in Uribia, located in La Guajira, which corresponds to a desert area according to the Lang's Index (LI) and with little or no surplus water according to the Thornthwaite climate classification. It is an approximation of what may happen to the abovementioned classifications under climate change scenarios.

As shown in **Table 2**, this study considered two models and two scenarios. The HadGem2-ES model was the optimistic scenario for the RCP 4.5 scenario in 2050 and the GDFL-CM3 model was the pessimistic scenario for the RCP 8.5 scenario in 2070.

A hydrological balance was made both for the baseline (1976–2050) and for the previously mentioned scenarios, based on the meteorological data measured at the Nazareth station and its corresponding future change scenarios due to the influence of climate change. **Table 3** shows the annual average values for each variable considered in the water balance.

The climate change models were reviewed, obtaining an increase of 1.7°C and a 2.4% decrease in precipitation for the optimistic scenario, as well as an increase of 3.7°C and an 11% decrease in precipitation for the pessimistic scenario. This will have a direct effect on the provisional crops, as soil moisture will be reduced.

As an example, **Table 4** summarizes the changes in the variables resulting from the water balance and the implications that these changes may have on the water requirements of the corn, bean, and melon crops in the study area.

Variable	Baseline value
Maximum temperature (°C)	32.5
Average temperature (°C)	27.3
Minimum temperature (°C)	22.0
Precipitation (mm)	510.2
Evaporation (mm)	2044.5
Evapotranspiration (mm)	1521.1

Table 3.
Meteorological variables of Nazareth station [6].

Water layer	Baseline	GDFL-ESM2G (GD) 4.52050			GDFL-ESM2G (HD) 6.02070			GDFL-ESM2G (HE) 8.52070		
		Value	Difference (mm)	Change (%)	Value	Difference (mm)	Change (%)	Value	Difference (mm)	Change (%)
Def (mm/year)	1060.6	1781.5	720.9	**68.0**	1816.1	755.5	**71.2**	1870.8	810.1	**76.4**
Pcp (mm/year)	510.2	446.2	−64.0	**−12.5**	445.2	−65	**−12.7**	446.2	−64.0	**−12.5**
ETo (mm/year)	1570.8	2227.7	656.9	**41.8**	2261.3	690.5	**44**	2317.0	746.1	**47.5**
ETa (mm/year)	510.2	446.2	−64	**−12.5**	445.2	−65	**−12.7**	446.2	−64	**−12.5**
Water requirements (mm/year)—Corn	257.5	426.9	169.4	**65.8**	426.2	168.7	**65.5**	433.7	176.2	**68.4**
Water requirements (mm/year)—Bean	195.9	363.1	167.2	**85.3**	356.6	160.7	**82.0**	370.2	174.3	**89.0**
Water requirements (mm/year)—Melon	223.9	403.8	179.9	**80.3**	394.1	170.2	**76.0**	412.5	188.6	**84.2**

The bold values highlight the percentage change.

Table 4.
Summary of water balance and requirements for the different scenarios and periods [14].

Once the water balance has been carried out with the new weather conditions, it is clear that none of the classifications mentioned at the beginning of this section have changed (Lang and Thornthwaite). However, the conditions of such classification are exacerbated since there is a reduction in the Moisture Index (Im), which may have a negative implication, both ecologically and socially, due to the lower availability and access to drinking water.

This suggests that there is a growing need to investigate this area in order to develop an adequate plan that may include water harvesting projects or efficient crop irrigation systems that take into account the future demands and projected precipitation deficits, to serve as a climate change adaptation strategy.

3.2 Effects of climate change on related activities

3.2.1 Sinú-Caribe basin

The influence of climate change in this area of the country was analyzed, evaluating the supply-demand relationship. The activities related to hydroelectric power were taken into account in four investigations that will be briefly described in this section.

This area of the country is deserted due to the anthropic activity in the riverbed of the basin, in addition to the deforestation associated with different activities. It is a region of great importance for the country considering that approximately 66% of the electricity in Colombia is generated in hydroelectric plants [15] and that 3.7% of that value is a direct contribution of the Urra I system, located in the Sinú-Caribe basin. Moreover, it is susceptible to flooding; so the occurrence of extreme precipitation events is important for research related to climate change, considering that due to its influence the events will tend to be more frequent in the future [3].

For the first investigation carried out in 2009 [7], the precipitation and temperature data were initially obtained for two weather stations in the area. Then, some statistical regressions were performed to establish the relationship between the mentioned variables and the flow measured in the basin. Regarding the future projection of precipitation and temperature, some scenarios were generated taking into account the data of climate change models (deltas) and the series observed in the basin stations.

When reviewing the precipitation projected by each of the models, it was concluded that some projects have an increase and others a decrease, having a possible percentage change between −21 and 14. However, the flow tends to decrease between 2 and 35% regardless of whether precipitation increases or not, given its relationship with other variables.

Given the above, the intention was to establish the sensitivity and vulnerability of the hydroelectric power generation system, creating scenarios based on the supply and demand of the area and the aforementioned climate projections. The scenarios were entered in the "Water Evaluation and Planning System-WEAP 2.1" and it was found that hydropower generation tends to decrease in the future, between 15 and 46%, which can increase the production costs that could be transferred to the users of this service. The above is summarized in **Table 5**.

This study indicates that research should be carried out in order to establish the potential vulnerability of systems related to supply and demand in this area under the influence of climate change. Therefore, a subsequent investigation was carried out in which it was considered that the impact on water resources was not only the result of the change in its offer or its quality, but also the pressure on the said offer in which the population and planning processes influence significantly [8].

Model/variable	Electricity generation		Flow
	Change %[1]	Change %[2]	Change %
CCSRNIES_A21	−0.7	−16.5	−5.9
CSIROMK2B_A21	−11.3	−25.4	−2.3
CGCM2_A21	−0.8	−16.5	−11.8
CGCM2_A22	−13.7	−27.4	−13.3
CGCM2_A23	−13.4	−27.1	−11.3
HadCM3_A21	−35.2	−45.5	−34.9
HadCM3_A22	−25.9	−37.7	−23.8
HadCM3_A23	−2.9	−18.3	−14.2
HadCM3_A2_SDSM	0.6	−15.4	−2.3
Statistic	−27.2	−38.8	

[1]*Reference scenario or baseline (1418.9 GWh/year).*
[2]*Maximum generation capacity (1687.2 GWh/year).*

Table 5.
Changes in hydroelectric power generation and its related variables [7].

A model was made using WEAP 2.1 software, to simulate hydrological processes in the scenarios concerning anthropogenic changes, land use, demands, and regulations among others. Its methodology is based on obtaining information on the daily consumption and activities in the study area, and their corresponding future increases based on projections. Then, it was contrasted with the reduction in supply due to environmental flow and quality considerations, as well as the possible variation in both temperature and precipitation and its consequent influence on the change in supply. Using WEAP, aspects of the supply-demand relationship were determined, such as unsatisfied demand, demand coverage, supply requirement, supply delivered, demand increase, and resource pressure indices. All scenarios are considered to be consistent with an increase in temperature and a positive or negative change in precipitation, as shown in Appendix 1, with a consequent decrease in the water flow or supply available.

As a result, it was concluded that in all scenarios it is not possible to supply the demand for water resources in all sectors and activities (commercial, industrial, rural, and urban areas), leading to a progressive growth of unsatisfied demand for the year 2039, as shown in **Figure 2**.

From this research, it is important to highlight the allocation of weights to climate change models based on their contribution to the flow, and their corresponding increase and decrease in temperature and precipitation respectively. Accordingly, a single weighted average scenario was proposed, where the worst possible scenarios are given a greater weight. Therefore, this scenario becomes a tool for planning water conservation strategies, such as the implementation of technologies for the reuse or efficient use of water, as well as the inclusion of vegetation conservation.

It was concluded that the proposed methodology and application of the WEAP tool allow to have a broader view of the resource in a region, allowing a better management of the water resource, with the possibility of planning strategies properly and taking into account adverse effects. This establishes the need to carry out a regionalization of climate change in order to develop tools that allow a detailed observation of its impact at local level. Finally, the inclusion of hydrological variables allows planning based on the water supply, reducing uncertainty in some way.

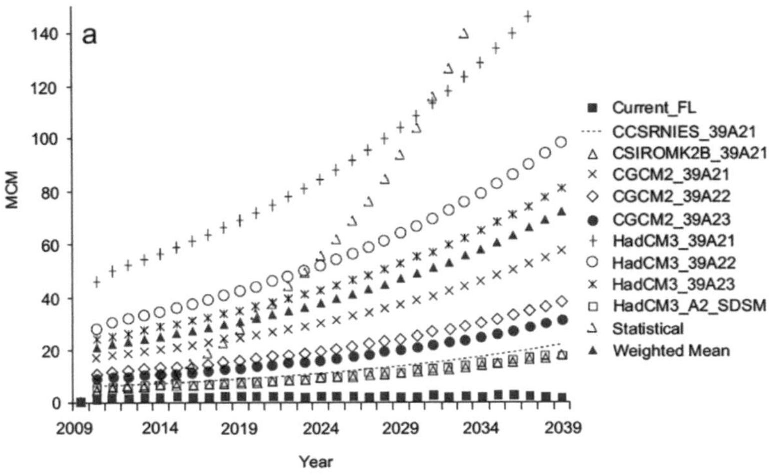

Figure 2.
Total unsatisfied demand, Sinú-Caribe basin [8].

Having reviewed the supply-demand relationship in this area for different activities and considering the importance of hydroelectric power generation for the country, the need to propose a vulnerability index for this activity was established [9]. Based on the observed and projected data from the previous investigations, both for precipitation and temperature and for flow, as well as on the supply and demand data of the water resource, the WEAP tool was used again to perform a water balance that includes the data mentioned and the operating policies, technical problems, bathymetries, and evaporation in the basin.

Given the above, there is a direct relationship between the flow and the volume of the basin with changes in precipitation and temperature. Therefore, the vulnerability index will depend directly on these two variables, its lowest value being the result of a 10% increase in precipitation and a 0.5°C increase in temperature and the highest value the result of a 10% decrease in precipitation and a 3°C increase in temperature.

Given the cost overruns in energy production, research similar to the previous one supposes the possibility for decision-makers to establish the vulnerability of their systems to climate change, as well as to implement projects to adapt to and mitigate the effects.

Finally, it was intended to establish if the availability of water can act as an optimization factor in the generation of hydroelectric power [10]; so, an investigation focused on this topic was carried out. To do so, using both current and future supply and demand values, an adapted scenario was established in which demand decreases by 20% due to the efficient use of water. This makes the critical point at which demand equals supply more distant, having a difference of 5 years on average.

As for hydroelectric generation, some system optimization scenarios were established, including pumping from a downstream point of the basin to an upstream point, which regulates the basin and can be used to mitigate events of flood that, as mentioned, are very likely to occur. The optimization scenarios are contingent on pumping with different start dates and the amount of water extracted. It was established that pumping must begin before the critical point established for each demand scenario to avoid regional conflicts over the use of water.

However, from this latest investigation it was concluded that the critical point with measures such as pumping only takes a little longer to occur. Therefore, it is crucial to make changes in demand such as establishing appropriate water management and regulation strategies, optimizing water delivery infrastructure, and establishing priorities. This study and the previous ones encourage decision-makers to carry out adaptation projects that take into account all the issues addressed in this section and the changes in the meteorological variables indicated in Appendix 1.

3.3 Effects of climate change on water resources, indices, and related activities

3.3.1 Nilo, Cundinamarca

The municipality of Nilo, located in the department of Cundinamarca, is an important region for the production of Cocoa in Colombia. This study seeks to evaluate the water requirements for growing this product in current and future scenarios with climate changes [11].

For this purpose, a baseline was initially established in the period 1975–2005. Then, using variables such as precipitation, temperature, evapotranspiration, among others, a water balance was performed to recognize and characterize the study area, establishing adequate water availability, according to the water resource indices in **Table 6**. This procedure was repeated again considering the change in the projected meteorological variables for the years 2050 and 2070.

Additionally, the water requirements of the crop were established using CropWat software, climatic variables of the baseline and future scenarios, as well as some parameters related to soil, which were established based on fieldwork performed in the study area.

Consequently, crops were delimited due to water deficiency in soil, as a result of an increase in temperature (T) and a decrease in precipitation (PCP). The aforementioned will involve drought stress, a possible increase in pests, and a drastic reduction in crop yield. In addition, there will be a possible increase in the water deficit (Def) in both the pessimistic and optimistic scenarios, changing the Hydric Availability Index (HAI) in the area from optimal to semiarid, as shown in **Table 7**. According to the values in **Table 7**, in terms of the water requirements of the reference crop, a value of 359 mm for the baseline and a consequent increase up to 535 mm were established as a result of climate change.

This study opens up the possibility of planning the use of the land, depending on the water requirements of both current and future crops, in order to make sustainable use of the water resource and can serve as a reference for new studies on this subject. This investigation measured the arithmetic average of the results obtained from the different models and scenarios. However, it does not allow observing the effect of each model, which may differ from each other, either in the magnitude of the change in temperature or in the increase or decrease in precipitation.

Index	Description	Value
LI	The Lang's index describes the humidity conditions in the area as the ratio between average annual precipitation and average annual temperature [16]	VH: Very humid H: Humid MH: Moderately humid ML: Semiarid L: Arid VL: Desert
HAI	The Hydric availability index allows to identify surpluses or deficiencies of water in specific areas or periods. It is a function of the relationship between the sum of the actual evapotranspiration and a quarter of the surpluses with the potential evapotranspiration [17]	VH: Very humid H: Humid MH: Moderately humid M: Optimal ML: Semiarid L: Arid VL: Desert

Table 6.
Description and assessment of the calculated indices.

Therefore, the scenarios were grouped into four clusters or groups of similar results, in which the centroid value of each of the variables was obtained, as shown in **Table 8**. It was decided to assemble a scenario by assigning weight factors to each cluster, in order to generate a unique scenario with the most adverse effects. Since the municipality of Nilo is mainly engaged in agriculture, it was concluded that the most negative effect in this area is the reduction of precipitation and the increase in temperature. Therefore, the clusters in which this occurs will have a greater value at the time of assigning the weight factor (WF) for each of variable [12].

Accordingly, a unique scenario (WA) for precipitation and average, minimum and maximum temperatures were established based on the previously assigned weights. This scenario differs from the arithmetic average (AA) of all the scenarios calculated; they were compared with the established baseline (BL), as shown in **Figure 3**.

Climate variable/scenario	**Tmin (°C)**	**Tmax (°C)**	**Tavg (°C)**	**PCP (mm)**	**Def (mm)**	**LI**	**HAI**
Current	21.4	23.4	22.4	1292.0	124.5	M-L	M
Change in climate variable/ scenario	***ΔTmin* (°C)**	***ΔTmax* (°C)**	***ΔTmed* (°C)**	**ΔPCP (%)**	**ΔDef (%)**	**LI**	**HAI**
Pessimistic 2050	2.0	2.1	2.1	−39.8	396.4	L	L
Optimistic 2050	1.3	2.3	1.8	3.2	31.1	M-L	M
Average 2050	1.6	2.1	1.8	−15.8	152.8	L	M-L
Pessimistic 2070	2.5	3.5	3.0	−36.5	382.8	L	L
Optimistic 2070	2.5	3.7	3.1	14.3	13.9	M-L	M
Average 2070	2.1	2.6	2.3	−14.0	150.1	L	M-L

Table 7.
Summary of the variation of climatic variables and associated indices [11].

Variable/cluster	**1**	**2**	**3**	**4**
Members	22	7	7	8
Percentage	50	15.91	15.91	18.18
Tmin (°C)	23.08	**24.61**	23.4	23.39
Tmax (°C)	33.89	**35.43**	34.06	34.1
Tavg (°C)	28.49	**30.03**	28.74	28.73
PCP (mm)	1101.25	1149.27	**874.74**	1314.08
Def (mm)	294.52	287.76	525.86	191.01
LI	L	L	L	M-L
HAI	M-L	M-L	M-L	M-L
PCP WF	0.26	0.68	**4.34**	0.38
Tmax WF	0.36	**2.46**	1.35	1.17
Tmin WF	0.34	**2.44**	1.39	1.23
Tavg WF	0.34	**2.45**	1.38	1.22

The bold values highlight the greater temperature values on the cluster 2 and so the greatest weighting factor for this variables correspond to the scenarios which are part of that cluster. While the minor precipitation value is in the cluster 3, giving it a greater weighting factor to this variable in the scenarios belonging to this cluster.

Table 8.
Summary of climatic variables, their associated indices and weights for each cluster [12].

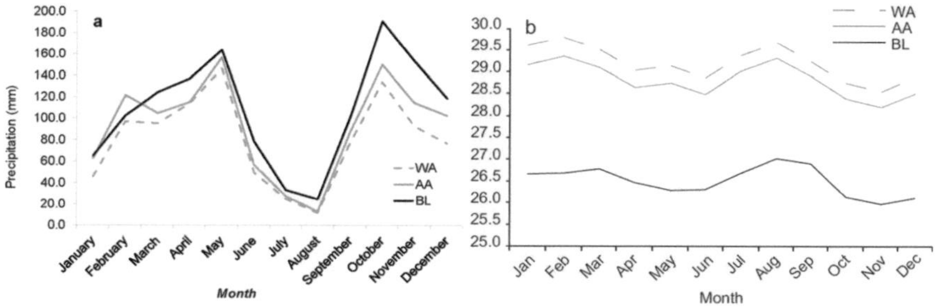

Figure 3.
Monthly behavior of (a) precipitation (b) average temperature [12].

Establishing a unique scenario as indicated in this investigation allows decision-makers to establish adaptation measures for that single scenario, focusing efforts on preventing or mitigating the most adverse effects for the area of interest, which is of vital importance for the proper management of water resources in an unfavorable future.

3.3.2 Coello River basin

The Coello River basin is located in the department of Tolima. It covers a large percentage of its territory and is of great importance for the region since it supplies the municipalities settled there, as well as their economic activities, mainly agricultural. Likewise, it supplies one of the most important irrigation districts in the region due to its large rice and cotton production.

An investigation was conducted in this basin, focused on assessing the implications of climate change on the supply, demand, and indicators of water resource status, namely, indices of aridity, water use, vulnerability to water shortage, and water retention. Its methodology was based on the development of a hydrological model using the Soil and Water Assessment Tool (SWAT), both for the baseline (1976–2005) and for the future period (2020–2050), in which daily precipitation and monthly temperature were entered as input variables.

In the case of temperature, the Delta Method was used as a methodology for the reduction of the geographic scale of the General Circulation Models (GCM) implemented, which consists in establishing the variation of the temperature per month taking as reference the historical data of the GCMs mentioned in **Table 2**. In the case of precipitation, since a daily resolution was required, it was decided to combine the Delta Method for the monthly scale reduction with the Maximum Entropy Method for the disaggregation of said value on a daily basis, depending on the observed behavior of said variable in each station studied [13].

As a result, a potential increase in annual precipitation was determined as shown in **Figure 4**. However, there was a sharp decrease in its value at daily resolution as illustrated in **Figure 5**, which suggests a possible increase in extreme events since large amounts of precipitation are concentrated on specific day(s).

In the case of the maximum temperature, its value increased progressively as illustrated in **Figure 6**.

Once the input variables for the hydrological model have been established, the flow values were obtained throughout the basin, thus establishing the water supply in each of the microbasins. Subsequently, the flows were characterized using flow duration curves, which, compared with the observed value, indicate an increase in the probability of extreme events and a decrease in the average flow that flows through the channel most of the time (**Figure 7**).

In the case of demand, having a projection of the meteorological variables and the crop areas, as well as the type of product grown, the water demand was established in terms of the irrigation requirement of said crops using the CropWat tool, as shown in **Table 9**.

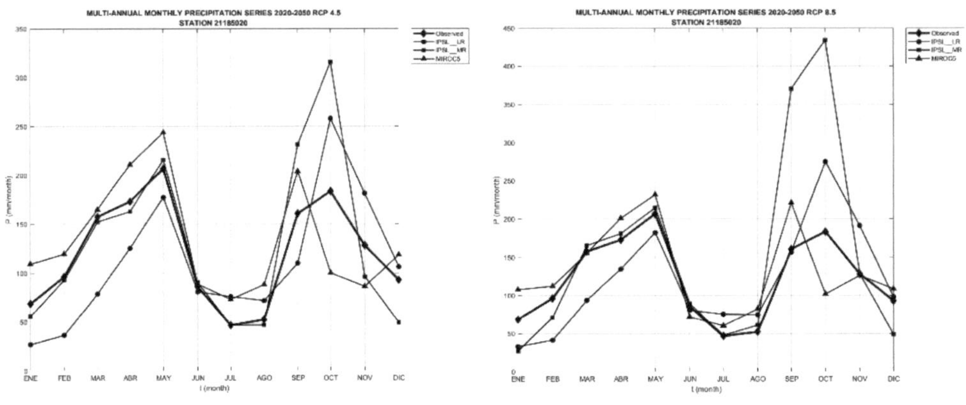

(a) (b)

Figure 4.
Monthly behavior of precipitation. (a) RCP 4.5 (b) RCP 8.5 [13].

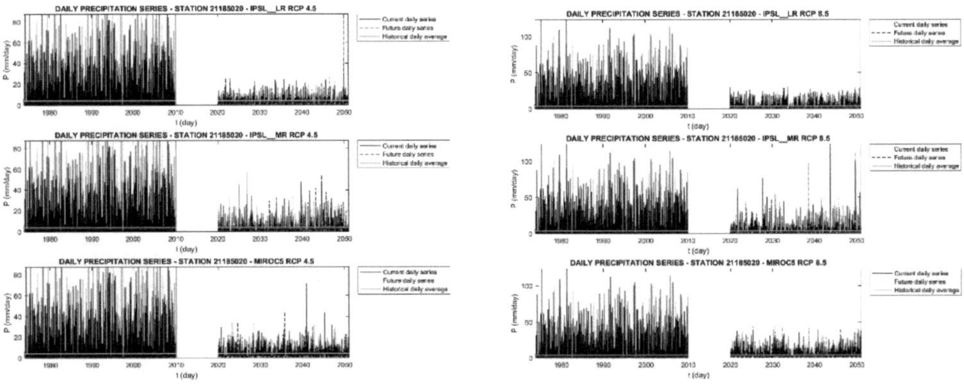

(a) (b)

Figure 5.
Daily behavior of precipitation. (a) RCP 4.5 (b) RCP 8.5 [13].

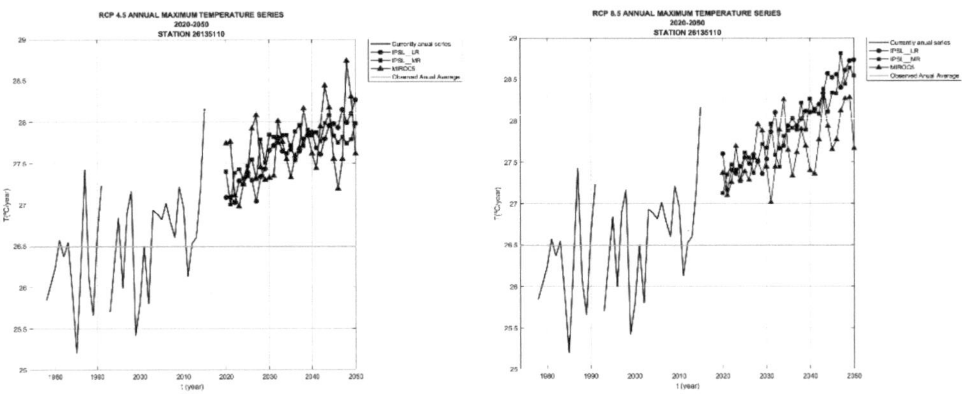

(a) (b)

Figure 6.
Annual behavior of temperature. (a) RCP 4.5 (b) RCP 8.5 [13].

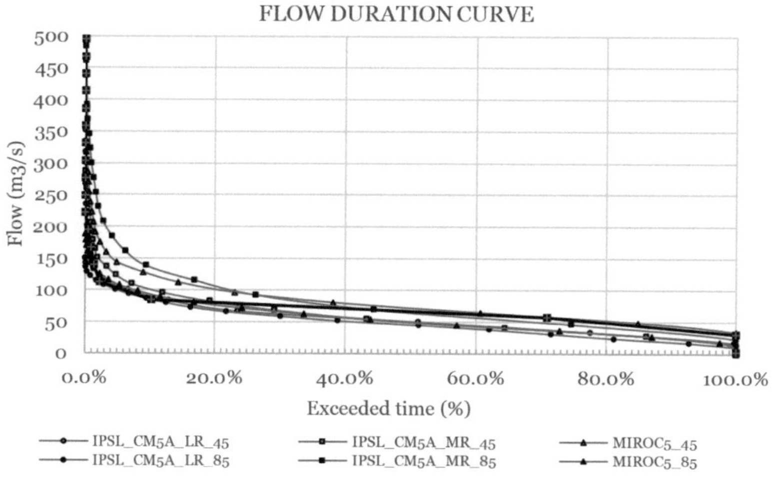

Figure 7.
Flow duration curve for all the models and RCPs selected, year 2050 [13].

Demand/crop	**Pane cane**	**Banana**	**Plantain**	**Arracacha**	**Rice**
Model	**(Mm^3/year)**	**(Mm^3/year)**	**(Mm^3/year)**	**(Mm^3/year)**	**(Mm^3/year)**
Current	13.5	10.6	5.4	2.0	296.3
IPSL_LR45	20.8	12.4	6.4	1.8	393.1
IPSL_MR45	16.3	14.2	7.3	2.2	367.5
MIROC5_45	20.4	10.3	5.3	1.7	314.4
IPSL_LR85	19.1	19.5	10.0	1.7	390.8
IPSL_MR85	21.2	14.1	7.2	2.2	362.9
MIROC5_85	13.1	10.2	5.2	1.9	322.2
Demand/crop	**Bean**	**Corn**	**Yucca**	**Cocoa**	**Coffee**
Model	**(Mm^3/year)**	**(Mm^3/year)**	**(Mm^3/year)**	**(Mm^3/year)**	**(Mm^3/year)**
Current	1.2	1.1	0.7	36.7	183.5
IPSL_LR45	1.6	1.3	0.98	50.7	245.5
IPSL_MR45	1.7	1.4	0.96	45.6	234.4
MIROC5_45	1.7	1.3	0.89	42.7	210.4
IPSL_LR85	1.5	1.3	0.94	49.8	243.6
IPSL_MR85	1.6	1.4	0.97	44	227.3
MIROC5_85	1.7	1.3	0.9	43.4	211.4
Demand/crop	**Mango**	**Avocado**	**Soursop**	**Lemon**	**Cotton**
Model	**(Mm^3/year)**	**(Mm^3/year)**	**(Mm^3/year)**	**(Mm^3/year)**	**(Mm^3/year)**
Current	2.6	0.042	0.014	0.33	12.3
IPSL_LR45	3.6	0.064	0.018	0.53	16.3
IPSL_MR45	3.2	0.054	0.018	0.47	14.7
MIROC5_45	3.0	0.042	0.018	0.38	12.2
IPSL_LR85	3.5	0.063	0.018	0.51	16.1
IPSL_MR85	3.1	0.055	0.017	0.47	14.8
MIROC5_85	3.09	0.046	0.018	0.39	13.1

Table 9.
Irrigation requirement for the crops in the study area [13].

Furthermore, growth projections were defined for both population and other economic activities such as livestock and industry. In this way, the supply that each of them will require in the future was established, noting that no demand for another activity is comparable with that established for the agricultural sector as shown in **Table 10**.

Finally, the water resource status indices were established, which are a function of both the previously calculated supply and demand. This evidenced a strong increase in the Aridity Index (AI) and Water Use Index (WUI), as well as a scenario of improvement with respect to the Index of Vulnerability to Water Shortages

Scenario	Agricultural sector demand (m^3/s)	Domestic sector demand (m^3/s)	Industrial sector demand (m^3/s)	Total demand (m^3/s)
Current	17.73	0.888	0.310	19
IPSL_CM5A LR RCP 4.5	27.55	1.579	1.790	31
IPSL_CM5A MR RCP 4.5	26.12	1.579	1.790	30
MIROC5 RCP 4.5	23.42	1.579	1.790	27
IPSL_CM5A LR RCP 8.5	27.67	1.579	1.790	31
IPSL_CM5A MR RCP 8.5	25.84	1.579	1.790	29
MIROC5 RCP 8.5	23.52	1.579	1.790	27

Table 10.
Water demand in the scenarios selected. Adapted from [13].

Index	Description	Value
AI	It describes the degree of surplus or deficiency of precipitation to sustain ecosystems based on potential and actual evapotranspiration, qualifying it from water deficit to water surplus [1]	VH: High surplus H: Surplus MH: Moderate to surplus M: Moderate ML: Moderate to deficient L: Deficit VL: Highly deficient
WRI	It establishes the ability to retain and regulate humidity in the basin, according to a relationship between the values extracted from the flow duration curve [13]	VH: Very high H: High M: Medium L: Low VL: Very low
WUI	It describes the pressure of the demand with respect to the supply [13]	VH: Very high H: High M: Medium L: Low VL: Very low
IVWS	It describes the vulnerability of the water system to the shortage of the resource for different users [1]	VH: Very high H: High M: Medium L: Low VL: Very low

Table 11.
Description and assessment of the calculated indices.

(IVWS), Water Retention Index (WRI), and/or the same indices calculated for the baseline. **Table 11** shows a brief explanation of these indices.

Based on the average values of each index, **Table 12** shows the total values of the basin.

Scenario	IA	WRI	WUI	IVWS
Baseline	M	H	H	M
IPSL-CM5A-LR – RCP 4.5	M	M	VH	H
IPSL-CM5A-MR – RCP 4.5	M	M	VH	H
MIROC 5 – RCP 4.5	M-H	M	VH	H
IPSL-CM5A-LR – RCP 8.5	M	H	VH	M
IPSL-CM5A-MR – RCP 8.5	M	M	VH	H
MIROC 5 – RCP 8.5	M-H	H	VH	M

Table 12.
Average values of each index for the baseline and each projected scenario. Adapted from [13].

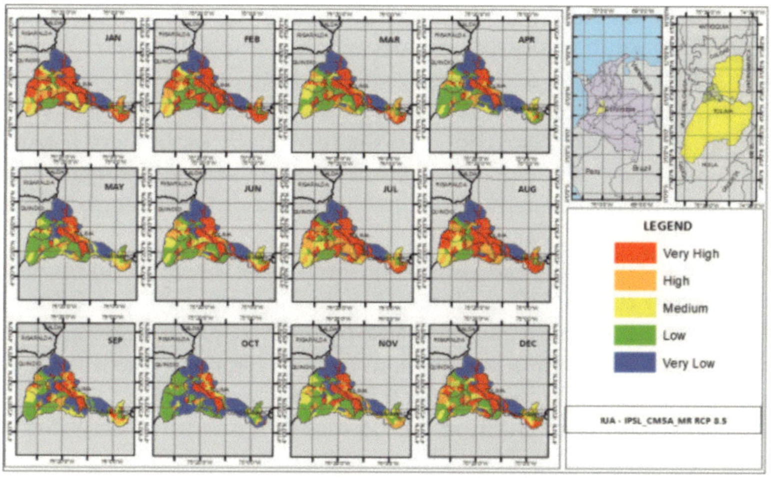

(a)

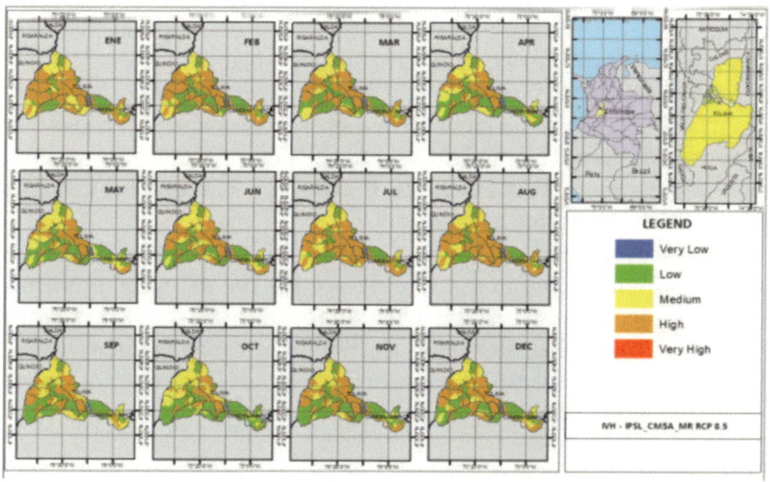

(b)

Figure 8.
Spatialized values for the IPSL-CM5A-MR RCP 8.5 model for (a) water use and (b) vulnerability to water shortages [13].

It is worth noting that these indices were calculated with a spatial distribution defined by the microbasins of the hydrological model, for each month of the year, so the research includes maps of each index such as those illustrated in **Figure 8**.

However, despite obtaining an optimistic future scenario unlike other regions of the country, this should be used with caution given the uncertainty that it entails; for example, there was an increase in the amount of monthly precipitation but a decrease in the daily average, indicating an increase in the intensities concentrated on specific days. The foregoing has a direct influence on the water retention index, which may give an unsuccessful perspective on the optimism of the scenario.

Finally, studies such as the one mentioned above may be of great interest to decision-makers, since they broaden their spectrum of possible future scenarios, in order to adopt measures that mitigate the possible impacts of climate change.

4. Conclusions

It was concluded that the effects of climate change throughout the country are very varied in contrast to its current state. These changes are summarized in Appendix 1, where there is mostly a decrease in the amount of precipitation in the future under climate change scenarios, with the exception of the Coello River basin area where there is an increment. It should be noted that in the latter case this increase occurs at monthly resolution and that on a daily scale there is a decrease in this variable with possible concentrations on specific days. This suggests a possible increase in precipitation intensity and consequently a possible increase in extreme events.

In the case of temperature, all studies agree that there will be an increase from 0.5 °C in the period 2011–2040 to almost 4 °C in the period 2071–2100 in different scenarios and areas as detailed in Appendix 1. The foregoing has a direct implication in related activities as explained in this chapter. A possible decrease in hydroelectric generation is expected given both the increase in temperature and the variation in precipitation. In addition, in the case of agricultural activities, the increase in temperature has a direct effect on evapotranspiration and consequently on the irrigation requirements of the crop, which will also depend on changes in precipitation.

In turn, the change in the variables, added to other anthropic activities expressed in terms of water demand, can exert significant pressure on the water resource. This could represent an increase in the vulnerability to shortages and unsatisfied demands, generating a risk associated with food security and water use for the population in the country. This, in addition to the potential risks associated with the increase in extreme events, that is, floods or droughts, has a direct impact on the inhabitants and their economic activities.

The latter stresses the need to conduct studies with a finer resolution both geographically and temporally in order to determine the potential impacts of climate change more accurately, serving as a tool for decision-makers. Using these tools, it is possible to establish strategies for the proper management of water resources, management plans that take into account future scenarios, as well as the importance of water availability to avoid regional conflicts.

Appendix 1

Changes in temperature and precipitation for the different studies in this chapter.

Site	Baseline		Scenarios	2011–2040		2041–2070		2071–2100		
	$\overline{T}$ (°C)	PCP_y (mm)		ΔT (°C)	ΔPCP_y (%)	ΔT (°C)	ΔPCP_y (%)	ΔT (°C)	ΔPCP_y (%)	
Nilo	26.5	1292	Arithmetic average			1.8	−16	2.3	−14	[11]
			Weighted average					2.7	−25.3	[12]
Uribia	27.3	510.2	HadGem – RCP 4.5			1.7	−2.4			[6]
			GDFL–CM3 - RCP 8.5					3.7	−11	
Sinú	28.2	2212	Had-CM3 - A2	1.3	7.5	2.0	19.7	2.5	30.4	[7]
			Had-CM3- B2	0.9	10.9	1.3	17.4	1.7	25.9	
			CCSRNIES_A21	0.5	−5.9					
			CSIROMK2B_A21	0.7	−2.3					
			CGCM2_A21	0.7	−11.8					
			CGCM2_A22	0.9	−13.3					
			CGCM2_A23	0.8	−11.3					
			HadCM3_A21	1.9	−34.9					
			HadCM3_A22	1.6	−23.8					
			HadCM3_A23	1.4	−14.2					
			HadCM3_A2_SDSM	0.5	−2.3					
Coello	18.8	1520	IPSL-CM5A-LR – RCP 4.5			1.1	−4.3			[13]
			IPSL-CM5A-MR – RCP 4.5			1.2	5.6			
			MIROC 5 – RCP 4.5			1.1	11.9			
			IPSL-CM5A-LR – RCP 8.5			1.4	2.5			
			IPSL-CM5A-MR – RCP 8.5			1.4	23.8			
			MIROC 5 – RCP 8.5			1.1	10.4			

Author details

Nathaly Güiza-Villa[1*], Carlos Gay-García[2] and Jesús Efren Ospina-Noreña[1]

1 Universidad Nacional de Colombia, Bogotá, Colombia

2 Research Program on Climate Change (PINCC), National Autonomous University of Mexico, Mexico

*Address all correspondence to: nguizav@unal.edu.co

References

[1] IDEAM. Estudio Nacional del Agua 2018. Bogotá: IDEAM; 2019. p. 452

[2] IDEAM, PNUD, MADS, DNP C. Nuevos escenarios de cambio climático para Colombia 2011-2100 Herramientas Científicas para la Toma de Decisiones – Enfoque Nacional – Departamental. Bogotá: Tercera Comunicación Nacional de Cambio Climático; 2015

[3] IPCC. Climate Change 2013 - The Physical Science Basis. In: Intergovernmental Panel on Climate Change. Climate Change 2013 - The Physical Science Basis. Cambridge: Cambridge University Press; 2014. pp. 1-30

[4] IPCC. In: Solomon S, Qin D, Manning M, Chen Z, Marquis M, Averyt KB, et al., editors. Climate Change 2007: The Physical Science Basis. Cambridge, United Kingdom and New York, NY, USA: Cambridge University Press; 2007. 996p

[5] Sanchez-Torres Esqueda G, Ospina-Noreña JE, Gay-García C, Conde C. Vulnerability of water resources to climate change scenarios. Impacts on the irrigation districts in the Guayalejo-Tamesí River basin, Tamaulipas, México. Atmosfera. 2011; **24**(1):141-155

[6] Ospina-Noreña JE, Domínguez-Ramírez CA, Vega-Rodríguez EE, Darghan-Contreras AE, Rodríguez-Molano LE. Analysis of the water balance under regional scenarios of climate change for arid zones of Colombia. Atmosfera. 2017;**30**(1): 63-76

[7] Ospina Noreña JE, Gay García C, Conde AC, Magaña VO, Sánchez Torres Esqueda G. Vulnerability of water resources in the face of potential climate change: Generation of hydroelectric power in Colombia. Atmosfera. 2009; **22**(3):229-252

[8] Ospina-Noreña JE, Gay-García C, Conde AC, Sánchez-Torres Esqueda G. Analysis of the water supply-demand relationship in the Sinú-Caribe basin, Colombia, under different climate change scenarios. Atmósfera. 2009; **22**(4):399-412

[9] Ospina Noreña JE, Gay García C, Conde AC, Sánchez Torres Esqueda G. A proposal for a vulnerability index for hydroelectricity generation in the face of potential climate change in Colombia. Atmosfera. 2011;**24**(3):329-346

[10] Ospina-Noreña JE, Gay-García C, Conde AC, Sánchez-Torres Esqueda G. Water availability as a limiting factor and optimization of hydropower generation as an adaptation strategy to climate change in the Sinú-Caribe river basin. Atmosfera. 2011;**24**(2):203-220

[11] Bermeo P. JEO-N. Evaluación de los requerimientos hídricos actuales y futuros, bajo escenarios de cambio climático en cultivos de cacao en el municipio de Nilo, Cundinamarca. Colombia. In: 2017 International Symposium on Cocoa Research (ISCR). Lima, Peru; 2017

[12] Ospina-Noreña JE, Bermeo Fúquene PA, Darghan Contreras E, Barrientos-Fuentes JC. Management and ensemble of future climate scenarios for specific agricultural systems in the municipality of Nilo, Cundinamarca (Colombia). Atmosfera. 2019;**32**(3): 197-211

[13] Güiza-Villa N. Estimación de los cambios en los índices asociados a la oferta y la demanda del recurso hídrico en la cuenca del Río Coello bajo escenarios de cambio climático [thesis]. Universidad Nacional de Colombia; 2019

[14] Ramírez D, Carlos A, Ramírez Soler CE, Ospina-Noreña JE. Análisis

de la disponibilidad de agua y requerimientos de riego bajo escenarios de cambio climático en La Guajira-Colombia. In: 6 to Congreso nacional de investigación en cambio climático; 2016

[15] UPME. Integración de las energías renovables no convencionales en Colombia. Bogotá D.C. 2015. p. 188. Available from: https://www.iamericas.org/documents/Renewables_in_Colombia.pdf

[16] IDEAM. Clasificación de los climas. 2009. Bogotá D.C. p. 18. Available from: http://atlas.ideam.gov.co/basefiles/clima-text.pdf

[17] IDEAM. Índice de disponibilidad hídrica (IDH), metodología de cálculo y aplicación en Colombia. Bogotá D.C.: Oficina del servicio de pronóstico y alertas; 2006

Chapter 7

The Art of Physical Hydraulic Modeling and Its Impact on the Water Resources of Pakistan

Muhammad Salik Javaid and Muhammad Zeshan Khalil

Abstract

Before any major hydraulic engineering project is undertaken for planning, designing, construction, or revamping and rehabilitation some kind of model study is but a necessity. Depending upon the time, resources and, significance of the project, the study could be done only on the paper and computer screen using some graphical, analytical or, statistical software and tools, or it could be combined with the more expensive and time consuming physical model study also. This chapter focuses on the question as to why the physical modeling should be reintroduced into engineering practice because of the modern techniques and systems now available for construction, operation and, data analysis of these physical models.

Keywords: hydraulic modeling, physical modeling, hydrological modeling, model, prototype, similitude, hydraulic research, Nandipur, barrage, weir, spillway, water resources

1. Introduction

The study of physical hydraulic models plays a role which is vital in the planning and designing of almost all hydraulic and hydrologic structures. May it be the stilling basins, spillways of barrages, river training works, hydraulic siphons, or even simple bridges, they are generally designed, evaluated, refined, and improved on the basis of physical hydraulic model studies. Physical model studies are comparatively expensive, costly, consume lots of time and resources to build and operate, and require technical labor and expertise in developing and testing the model. The selection of appropriate scale ratios between prototype and model plays a very significant and imperative role for the reliability and rationality of the results obtained.

Researchers and engineers working in the field, face a real challenge once they have to finalize on the basis of physical and/or numerical models, the rehabilitation and modernization works for any already constructed and operational hydraulic structure. The success of any rehabilitation work depends upon the precise and accurate identification of hydraulic and hydrologic problems on the prototype structure, because any failure may lead to partial or complete wastage of huge investments.

2. Hydraulic modeling basics

The laws of similitude enable a researcher to predict the likely performance of prototype hydraulic structures from tests made with far less expensive models. We need not use the same fluid for the model as the prototype. We may obtain valuable results at a minimum cost from the tests conducted on the small scale hydraulic models. Any textbook on hydraulic physical modeling will tell us that the following similarities have to be ensured between the model and the prototype hydraulic structure [1, 2].

2.1 Geometric similarity

Model and prototype should have identical shapes but differ only in size as per the defined scale ratio. This would ensure geometrically similar flows. Under certain conditions, distorted models are resorted to by having different scale ratios for the lateral, longitudinal, and vertical directions, but then the same has to be incorporated during the interpretation of results.

2.2 Kinematic similarity

Ratios of the velocities on all corresponding points on the model and prototype hydraulic structure should be the same to ensure the same kinematics of flow.

2.3 Dynamic similarity

The quantum and direction of all forces acting on the corresponding points on the model and prototype should be in the same ratio, to ensure the same dynamics of the flow. Dynamic similarity can also be ensured by ensuring similarity of the combination of forces, by following the Froude Law, Reynolds Law, Mach Law, etc., for modeling.

3. Physical hydraulic modeling

Physical modeling of hydraulic structures has been in use since the times of Leonardo Da Vinci. However, since then this art and science have gone manifold changes, developments, and positive improvements. Such models provide a visual insight into the hydraulic phenomena of water and fluid flows. These models also provide technical flow data through the elaborate system of instrumentation provided. The data and flow visuals can be recorded for future reference, computations, training materials, and records.

The role of hydrological modeling has been well described in [3], wherein the authors reiterate that hydrological models are in fact basic, theoretical, and physical representations of the hydrologic cycle, and these are often used for the understanding and prediction of hydrological processes. They categorize the hydrological models as (a) models which are based on data collection, and (b) black-box models which are based on process description.

Because of the importance and special role of physical hydraulic modeling, various renowned organizations have developed their physical hydraulic research centers. The most common and well-known are the Waterways Experiment Station (WES) of the US Army Corps of Engineers and Hydraulic Research Station (HRS) of Punjab Irrigation Department, Pakistan.

3.1 Waterways Experiment Station (WES)

The US Army Corps of Engineers Waterways Experiment Station (WES) was created in 1929 to provide support for the vast flood control plan for the entire lower Mississippi valley after the tragedy of the 1927 most horrific river flood. The WES laboratory complex located at Vicksburg, Mississippi is now the principle research, testing, and development facility, which supports studies in many other fields in addition to its primary field of hydraulic engineering. WES provides services for training, and technical assistance, research, and also software development, which reflects the state-of-the-art expertise of WES in hydrologic engineering and closely associated fields of planning analysis. In its research and development work, WES uses more application of model experiments employing the principles of hydraulics. WES has made a significant contribution through the publication and distribution of its research reports.

3.2 Hydraulic Research Station (HRS)

Hydraulic Research Station, located at Nandipur near Gujranwala, in Pakistan is one of the largest research laboratories in the world. This field research station was established in 1926 and is under the administrative control of the Irrigation Research Institute, Lahore being its field station. The Nandipur station has 40 hectares of land divided into 22 research bays commonly called as research trays. Through a small irrigation channel, the water availability of 15 cumecs and a gravity head of 4 meters is provided, however for higher heads pumping facility is also available. The Nandipur Hydraulic Research Station meets the requirement of the study of numerous problems that are related to planning, operation, and management of water resources. Physical models for almost all the major irrigation and hydraulic structures now present in the country have been run, tested, and optimized at this station.

Hydraulic Research Station at Nandipur has carried out model studies of almost all major hydraulic engineering projects undertaken in Pakistan and India in the pre-partition as well as the post-partition era. The major projects of Mangla Dam and Tarbela Dam which were constructed as part of the Indus Basin Treaty were also modeled in this facility. Many other barrages, weirs, link canals, and river training works have been modeled and approved prior to the finalization of their designs. A sample of the physical hydraulic modeling projects undertaken by the Hydraulic Research Station is displayed in **Figures 1** and 2.

Figure 1.
Flow from Flip Bucket Energy Dissipater.

Figure 2.
Model of a Typical Barrage.

In the recent past, the rationality of the massive hydraulic structure of Jinnah Barrage [4, 5] was questioned as a model study indicated that at existing conditions of water levels the formed hydraulic jump was located on the glacis only up to a discharge of 400,000 cusecs. The hydraulic performance of the barrage, under-sluices, silt excluders, and also the subsidiary weir was yet not tested at higher discharges. Mahboob [6, 7] reviewed the design of Kalabagh Barrage and he found it acceptable only after the physical hydraulic model study because the hydraulic modeling study for energy dissipation under the conditions of existing water levels pointed out that hydraulic jump over the horizontal floor was repelled by the excessive lowering of the channel bed at the downstream (retrogression) (**Figure 3**).

Figure 3.
Model Study of Taunsa Hydro Power Project.

4. Physical modeling: a case study

The hydraulic modeling study cited here targets to examine sedimentation aspects of two cascade reservoirs on Poonch River; with the help of physical modeling and numerical simulation. A physical model of Poonch River was prepared at Nandipur Research Institute to study the sediment transport behavior [8]. After the base test, the model was used to get data for various scenarios of sediment flushing in the cascade reservoir system. The River geometry, riverbanks, hydraulic structures, cross-sections, and other physical attributes of the river were prepared from a topographic

Figure 4.
Model Study of Poonch River Sedimentation Project.

Figure 5.
Model Study of Poonch River Sedimentation Project.

survey using AutoCAD. These files were used in HEC-RAS and BASEMENT for simulations (**Figure 4**).

Delta profile and flushing were modeled by HEC-RAS 5.0. The simulation showed that the life of the un-sluiced Gulpur HPP is about 14–15 years and that of Rajdhani is about 35 years. To enhance the life of the project, annually 4–5 days are required for flushing with an optimized discharge of about 250 m^3/s. Model verification was performed by calculating the bed topography and flushing efficiency. The results obtained through the model were consistent with bed changes, demonstrating its suitability for the regeneration of regression channels and lateral erosion (**Figure 5**).

5. Revival of the vanishing art

Other techniques in addition to physical hydraulic modeling available to a researcher are mathematical modeling, statistical modeling, and numerical modeling. With the advent of modern computers having speedy and fast processors, massive data storage, better data management software, and intelligent computational techniques the statistical modeling and numerical modeling have become the favorites of every researcher and engineer. The cutting edge graphics cards and attractive presentation techniques have also added to the magnetism of such indoor modeling. However, despite all this, the value and importance of physical hydraulic modeling cannot be overshadowed by these. The natural intricacies, physical behavior, the kinematics and dynamics of all fluids and especially large mass flows of water can only be studied through physical modeling.

With the innovation of new materials of construction including the nanomaterials, the physical hydraulic modeling has been revived. Now very intricate designs can be created and manufactured using new and modern materials. The same is true for hybrid and very strong epoxies and sealing materials which now help in making watertight models. Fabrication of models and their miniature parts has also been revolutionized by laser cutting, computerized numerical machines that can make precision model parts.

Revolution in measuring instruments for all hydraulic parameters has also provided a quantum jump to physical hydraulic modeling. Doppler velocimetry, very sensitive and accurate probes and pressure transducers, laser leveling gauges, and other such instrumentation can now be used to obtain and collect very sophisticated data for physical hydraulic models.

The latest techniques in flow visualization have done wonders in fluid mechanics and hydraulic modeling. Modern electronics and advancement in graphics, optics, and sensors has revitalized the hydraulic modeling and made it an advanced and modern field of science and technology.

On the other hand, the models based on process description also called deterministic models are rather complicated as compared to the stochastic hydrological models representing surface runoff, channel flow, subsurface flow, and evapotranspiration. Such models cannot by physically modeled, and therefore these have to be computer modeled [3].

6. Recommendations and conclusion

The art, science, and technique of planning, construction, and operation of physical hydraulic modeling are losing the race against numerical and computer modeling. However, there is a dire need that due to its very special place in research and investigation, this modeling technique should remain in vogue. For this very purpose its education, teaching, and engineering practice may be included in the curricula of various universities, colleges, and other technical training institutes.

For very important and significant hydraulic structures, the failure of which cannot be afforded due to various reasons, it may be made mandatory that physical hydraulic modeling is carried out prior to the finalization of designs of construction and rehabilitation.

Author details

Muhammad Salik Javaid[1*] and Muhammad Zeshan Khalil[2]

1 Morgan State University, Baltimore, MD, USA

2 Civil Engineering Faculty, Abasyn University, Islamabad, Pakistan

*Address all correspondence to: msalikj@morgan.edu

References

[1] Kaleem MK. Fluid Mechanics, and Machinery. India: Oxford University Press; 2015

[2] Franzini, J.B. and Finnemore, E.J., Fluid Mechanics with Engineering Applications. McGraw-Hill Education. October 2001.

[3] Javaid MS, Khalid L. “Hydrologic Science - Where we stand today?” In: “Hydrology – The Science of Water”. London, UK: IntechOpen Publishers; September 2019. pp. p1-p6

[4] Chaudhry AZ. Energy Dissipation Problems Downstream of the Jinnah Barrage. Pakistan Journal of Engineering and Applied Sciences. 2008;**3**:19-25

[5] Chaudhry AZ, Nasim NH, Shakoor A. Alternative Report to Subsidiary Weir, Rehabilitation and Modernization of Jinnah Barrage Project. Government of Punjab, Pakistan: Irrigation and Power Department; April 2008

[6] Mahboob, S. I. Design, and Construction of Kalabagh Barrage, Punjab Engineering Congress, Paper No. 34, 1943.

[7] Mahboob S. I. The Kalabagh Barrage, Punjab Engineering Congress, Paper No. 251, 1942.

[8] Munawar Iqbal, Modeling for Optimization of Parameters of Sediment Transport for Cascade Hydropower Projects on Poonch River, doctoral thesis, UET Taxila, 2019.

Chapter 8

Water Resources and Their Management in an Increasing Urban Demography: The Case of Dakar City in Senegal

Cheikh Faye

Abstract

The United Nations classifies Senegal as a water-poor country (less than 1000 m^3 per capita of freshwater reserves) and about 20% of its population did not have access to a drinking water supply (estimates of 2015). Economic growth and the fight against poverty in Senegal depend essentially on the availability of water for the development of agricultural and industrial activities, in addition to satisfying domestic uses. As a developing country, Senegal's human, monetary and institutional capacities are often limited to providing clean and sufficient water efficiently to its citizens. This article examines the management of water scarcity in the city of Dakar (capital of Senegal) in a context of increasing demography and urbanization. However, Senegal has sufficient water resources to meet the demand if the available resources are properly managed. As a result, several initiatives are under way in Senegal to mitigate water problems and protect the country's water resources: reducing pollution, improving access to drinking water and setting up rational and equitable exploitation with a constant concern for sustainable development.

Keywords: water resources, management, urbanization, water scarcity, water policy

1. Introduction

The Republic of Senegal, located at the extreme western tip of the African continent, covers an area of 196,722 km^2 for a population of 13,508,715 inhabitants [1]. There are three climatic domains in Senegal, from south to north: the southern Sudan, northern Sudan and Sahelian domains, each domain having two variants (coastal and continental) [2]. Located in the tropical zone, Senegal has a Sudano-Sahelian climate with annual rainfall ranging from about 1250 mm in the south to just over 200 mm in the north. The potential of Senegal's water resources (surface and groundwater) is important. Three rivers originating from Guinea (the Senegal, Gambia and Kayanga rivers) irrigate a large part of the country [3, 4]. Alongside these two large rivers, there are smaller rivers characterized by intermittent flows (Casamance, Kayanga, Sine-Saloum (**Figure 1**). Five management and planning units (PMUs) have been established for the management of these different categories of water resources (1. Senegal River Valley, 2. Peanut Basin, 3. Senegal-East, 4. Casamance, 5. Cape Verde Peninsula) subdivided into 28 sub-units (**Figure 1**) [5].

Figure 1.
Water resources management and planning unit in Senegal (source: DGPRE).

Groundwater is also an essential component of Senegal's water potential and generally consists of four major aquifer systems corresponding to the main geological formations: the superficial aquifer system or "terminal complex" (Quaternary); the intermediate aquifer system (Eocene and Paleocene); the deep aquifer system (or Maestrichtian); the aquifer system of the basement [6].

In Senegal, the potential for water resources (surface and groundwater) is high and the availability of renewable water is currently estimated at around 4747 m^3/inhabitant/year [6]. Estimates indicate about 80% of its population have access to a drinking water supply in 2915 [7]. However, the United Nations classifies Senegal as a water-poor country with less than 1000 m^3 per capita [8]. Thus, the issue of water has become a national concern given the range of issues facing the sector [9]. These problems include, among others, climate variability, vulnerability of water resources, poor distribution of water availability in space and time, poor water quality in some places. The water crisis can be explained both by the absolute lack of physical availability, poverty and inadequate water management policies. In general, Senegal has a large potential for water resources, but its uneven distribution, its overabundance in the rainy season often causes catastrophic floods and shortage in the dry season causes severe drought conditions resulting in crop losses, livestock, public health problems and environmental degradation [10].

These numerous factors, such as global warming (recurring and severe droughts and floods), contamination of drinking water and lack of investment in water resources have exacerbated the water crisis, whose role in the achievement of its development objectives is incommensurate [11]. Its economic performance and the reduction of poverty depend mainly on the availability of drinking water. A set of economic activities in Senegal (agriculture, industrialization, energy production and tourism) are inherent to the availability of water resources. At the same time, access to safe and sufficient water is necessary for the well-being of the population.

Senegal is home to some major cities, namely Dakar, Pikine and Touba, Thies. The capital of Senegal, Dakar, was founded by Faidherbe in 1862, on the site of a fishing village. It was the capital of the AOF from 1902. Enjoying a strategic geographical location, the city is since the colonial era, a maritime and air junction between Africa, Europe and America. It covers an area of 550 km^2 and has about

23% of the total population of Senegal estimated 15,256,346 inhabitants, according to demographic projections in 2017 [12]. It is the largest city in Senegal and is its political, administrative, economic and cultural center.

The water cuts that have become commonplace in many parts of the country, affect more seriously the capital, Dakar where the daily deficit is estimated at more than 20,000 m^3. With a consumption of 360,000 m^3/day, the capital struggles to quench its thirst and satisfy its water needs. The Senegalese Water (SDE) needs 300,000 m^3 per day to meet the demand of the Dakar population, but it drags a heavy deficit in the correct supply of water. At present, a large part of the city's population does not have access to running water 24 hours a day. From 200,000 subscribers in 1996, the SDE is now struggling to satisfy its 800,000 current subscribers.

In Senegal, the urban population is estimated at 6,541,504 people in 2015 including 3,360,728 for the Dakar region [12]. In these urban areas, the rate of access by connection within the covered perimeter stabilizes in December 2015 at 88.9% with 96.2% for the Dakar region. Thus the total production of water was 172.27 million m^3 against an annual forecast of 169.7 million m^3, or 110% of realization. For the Dakar water supply system (WSS), production reached 124.2 million m^3 [13]. As a result, there is a tendency to saturate the facilities of the Dakar AEP (for example, Mékhé is running at almost 24 hours). Because of the size of the population of Dakar, part of this population does not have access to either an improved water source or running water. For example, some outlying districts (in the suburbs of Dakar) do not receive water every day, while others almost never receive water because of the low water pressure.

In some cities in Senegal, the biggest challenge in the city is often lack of water supply. For example, residents obtain water from individual connections, public connections, wells, springs and water vendors that are not monitored [11]. In addition, water pollution does not save tap water and its consumption can often be harmful to the health of populations. It is for this reason that a good part of the population prefer bottled water for their consumption.

The city of Dakar has been chosen for the management of water scarcity in a context of growing demography and urbanization. Due to increased water demand in the Senegalese capital, residents of some neighborhoods are frequently randomly supplied with running water, with the suburbs being the most affected. The ever increasing hydraulic equipment fails to meet the needs of the population. The difficulty of supplying drinking water, which often strikes the Dakar inhabitants, comes in the context of global warming and the removal of resources that are increasingly important. Suddenly, the Senegalese government must expand its production capacity and distribution networks if it wants to keep pace with unprecedented population growth.

2. The challenges facing the water sector in Senegal's main cities

As a developing country, Senegal's human, monetary and institutional capacities are often limited to providing clean and sufficient water efficiently to its citizens. Water scarcity in large cities like Dakar is chronic and continues to worsen with increasing urbanization and pollution of the resource. Most of the time, SDE water disruption notices are the norm in urban areas (**Figure 2**). Water scarcity may worsen in the future for several reasons: (1) increased water demand due to rapid urbanization, (2) poor water management, (3) degradation continuous water sources, (4) irregular weather conditions, (5) old and dilapidated water infrastructure, and (6) the incompetence of the water distribution companies.

2.1 Growing populations and urbanization

Senegal, like many other developing countries in Africa, is experiencing rapid urbanization. In 1960, its population was estimated at 3.207 million people and in 2017, 15.851 million people (**Figure 3**) [14]. The urban population has grown from 738,000 in 1960 to 7409 million in 2017 (**Figure 4**). This urban population, which accounted for only 23% of the total population of the country in 1960, therefore rose to 45.9% in 2015. The share of the urban population increased considerably between 1960 and 2014. The data show that 23% of the population living in urban areas in 1960 increased to 45.9% in 2015. The share of the population living in urban areas has increased dramatically in Senegal in recent decades. Thus, the urban growth rate is 4.78 and 3.46%, respectively over the periods 1960–1985 and

SDE
SÉNÉGALAISE DES EAUX

The Senegalese Waters informs its customers that because of the repair work of the main line of Lake Guers (ALG1) caused by the French company SOGEA / SATOM working on behalf of SONES in Ndande this Monday, April 08 2019, some of Dakar's water supply system production works are shut down. The water distribution will thus experience disturbances ranging from the decrease in pressure to the lack of water in the following areas:
- The localities supplied by the conduits of Lake Guiers in the regions of Louga and Thiès;
- Rufisque and surroundings;
- Dakar and its suburbs.

A tanker truck will be put in place to relieve the populations of the neighborhoods most affected by these disturbances.
The situation will gradually return to normal at the end of the work scheduled on Tuesday, April 09, 2019 in the evening.
The Senegalese Waters apologizes to customers for these inconveniences beyond the control.
For more information, contact 800 00 11 11 (toll free).

Figure 2.
Notice of disturbance of water supply by Senegalese water.

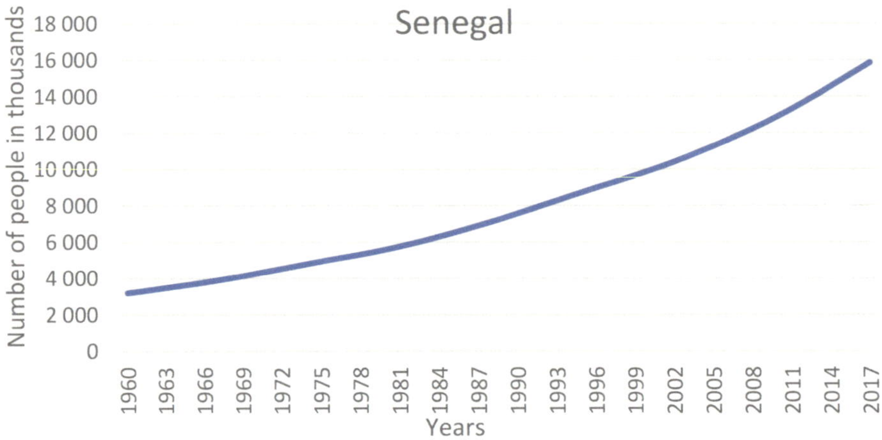

Figure 3.
Senegal's total population: semi-annual estimates of the resident population from 1960 to 2017 (source: World Bank database).

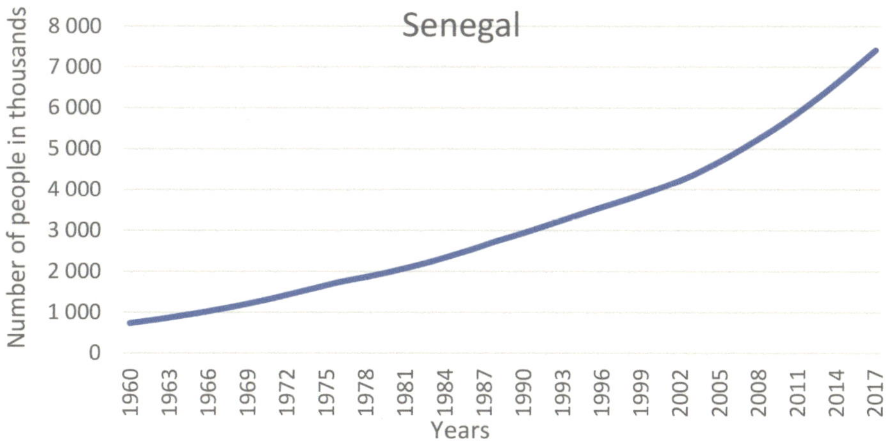

Figure 4.
Urban population of Senegal: semi-annual estimates of the resident population from 1960 to 2017 (source: World Bank database).

1985–2015. This rapid pace of urban growth can be explained by the drought of the 1970s [2] and its corollary rural exodus.

According to United Nations estimates, the urban population of Senegal will increase to 11.778 million by 2030, which will represent 53.24% of the national population (22.123 million in 2030). As for the city of Dakar, its population will increase to 4.339 million by 2030 (**Figure 5**), which will represent more than 38.84% of the urban population [14]. Urbanization does not only mean swelling of the population, but also an increase in the area requiring better services. The rural exodus from sub-Saharan Africa is the main cause of the rapid growth of the urban population. To cope with the strong urban growth, water production has risen drastically, from 95.32 million m^3 in 1997 to 172.27 million m^3 in 2015 (**Figure 6**). The growing population (**Figures 4** and **5**) continues to put pressure on available water resources, resulting in a reduction in per capita water availability. According to projections, 70% of the world's population will live in urban areas by 2050 [15]. The growing population continues to increase the demand for water for domestic, industrial and agricultural purposes.

Africa currently has three main causes: (1) displacement of people from rural areas to urban centers; (2) the increase in the urban population, especially when economic opportunities extend to previously rural areas; and (3) the development of previously rural areas in urban areas due to increased economic activity [11]. Although it is difficult to define exactly an urban environment, it is widely accepted that the term could be determined by location, size of population and percentage of non-agricultural activities, pressures on environmental resources such as water. In recent years, urbanization of rural areas has increased considerably.

Although urbanization has many economic and social benefits, it causes many environmental problems such as loss of biodiversity, air and water pollution and increased pressure on arable land [16]. It has directly affected the availability and quality of water due to increased demand and pollution resulting from its many applications. Many developing countries are facing the problem of access to safe drinking water. In Senegal, rapid urbanization has prevented some cities from coping with the huge demand for clean and sufficient water. In Dakar, faced with strong urban growth (an average of 120,000 people per year), the improvement and expansion of infrastructure is very expensive and, as a result, does not generally keep pace with the growth. This makes wastewater management very problematic.

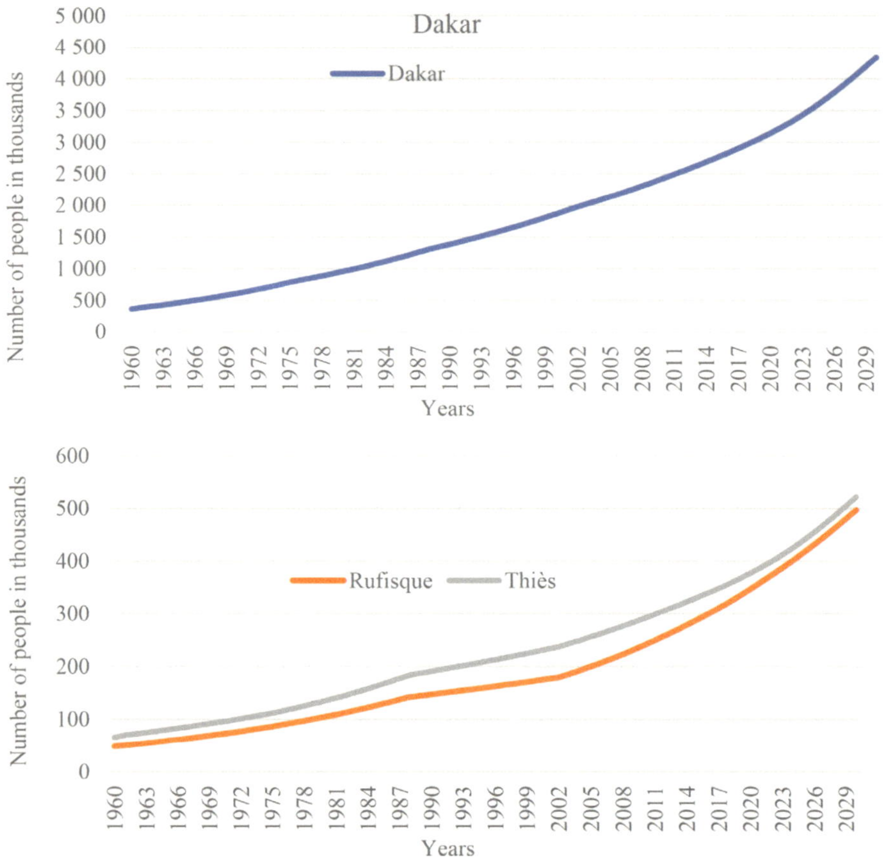

Figure 5.
Urban population of three cities in Senegal: semi-annual estimates of the resident population from 1960 to 2035 (source: World Bank database).

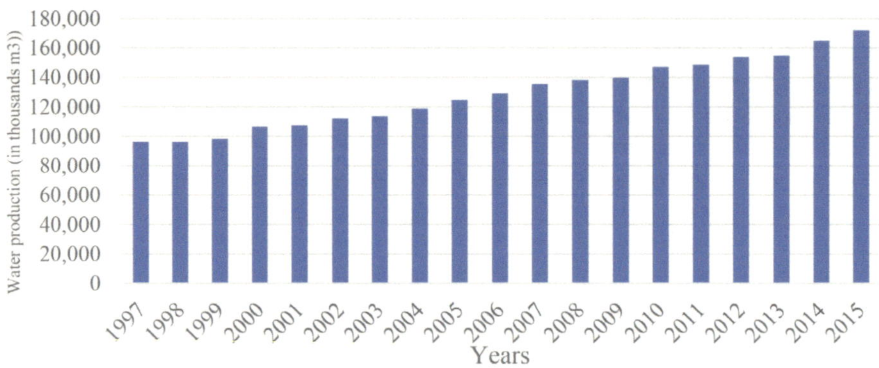

Figure 6.
Evolution of water production by the SDE (SONES perimeter) from 1997 to 2015.

In addition, pollution from agricultural production (urban farmers irrigating their crops with untreated wastewater) and industrial production has become one of the biggest challenges for Senegal's water resources [17]. In addition, because of the growing population and its corollary the increase in water needs, the lack of protection of water sources, the scarcity of the resource and the pollution it faces are a source of hindrance improved and protected water.

DOI: http://dx.doi.org/10.5772/intechopen.90218

2.2 Contamination of available water

The water sources available in Senegal are frequently affected by pollution of chemical, microbiological or thermal origin. Chemical contamination of this water, often used for drinking, can result from the presence of excess nutrients, acidification, salinity, heavy metals and organic pollutants [18]. Reports indicate that industries at 32.5% and agriculture at 14% are the sectors that contribute most to the economic development of any population [19]. On the other hand, 80% of the water contamination comes from these two important sectors. Agricultural practices, industrialization, mining, and open sewer lines parallel to the water system are responsible for most of the problems affecting water quality (**Figure 7**).

The lack of adequate management of liquid and solid waste results in the deposit of this waste directly into water bodies (**Figure 7**), which contributes to the vicious circle of water destruction. In fact, the growth and development of agriculture in Senegal has led to an increase in the use of fertilizers. Agrochemicals end up in bodies of water causing considerable pollution. In addition, most industrial water treatment plants discharge partially treated or totally untreated effluents into surface water sources, which often contain high levels of toxic substances. These pollutants and other pollutants of domestic origin continue to cause environmental problems [11]. Many Senegalese living in informal urban areas lack access to safe drinking water, often resulting in multiple epidemics that affect their health and livelihoods. In addition, large leaks in water pipes (**Figure 7**), dilapidated infrastructure and illegal connections still hinder the availability of drinking water supply. Due to the large leaks in the water channels, the treated water is sometimes contaminated before reaching the users.

Figure 7.
Water quality degradation factors in Senegal: (1) CSS effluent discharges into Lake Guiers; (2) water line between the lake and Dakar damaged; (3) domestic uses on Falémé; (4) equipment for washing gold on the banks of the Falémé.

2.3 Degradation of water sources

The main sources of water in Senegal, beyond the groundwater, are surface water (rivers and lakes that are very attractive for agriculture and populations). The watersheds that cross the national territory have experienced two major pressures in recent years on their water resources: (a) pressures from natural sources (climate variability and change); (b) anthropogenic pressures (dams, rapid population growth and various productive activities) [3]. These pressures have had repercussions on the natural environment of the basin and its ecological diversity [20]. They have resulted in watershed degradation that has diverse and unpleasant consequences, often resulting in increased runoff, flash floods, reduced infiltration, erosion and siltation, to name just a few examples. The impacts of human activities (development, rapid population growth and various productive activities) related to the exploitation of resources for the satisfaction of the daily needs of the populations sometimes manifest themselves negatively on all the natural resources of the basin. Environmental protection of the watershed is important for the safety and sustainability of urban water supply. A healthy ecosystem ensures quality water for cities, reducing treatment costs and the danger to human health.

Activities such as gold mining are causing degradation of natural resources in the basin through reduced vegetation cover and deterioration of water quality. With the use of chemicals in the practice of this activity (leaching technique with cyanide or mercury and tailings ponds), pollution of water resources in the basin is multiplying. With the major developments (Diama and Manantali dams), flow control and water permanence in the basin have led to the partitioning of mining activities (extraction and washing with mercury release) and agricultural activities (large consumer fertilizers and pesticides) with far-reaching consequences [21]. This results in severe degradation and deterioration of animal and plant resources [22]. With this destruction of freshwater ecosystems, the Senegal River finds itself in a situation of loss of some of its functions, how important, and may alter its hydrological functioning [23].

2.4 Invasive species

Biotic factors affect water resources. For example, the presence of invasive species such as hyacinth, Salvinia, Pistia and Typha causes ecological imbalance. The proliferation of plants and invasive species in major bodies of water, including aquatic plants, is of increasing concern. Typha and algae are plants that contribute to the deterioration of the water quality of the lake because of the very toxic substances secreted that can reduce the good water quality (**Figure 8**). They degrade surface water resources and may even have contributed to the eutrophication of freshwater lake ecosystems. These invasive species block rivers and greatly influence water quality. In Senegal, built dams (such as Diama and Manantali on the Senegal River) play an important role in the reliable and sustainable supply of water. However, these developments, by permitting the permanence and softening of the water, have led to the proliferation of certain species such as Typha australis and *Pistia stratiotes* and the appearance of new species such as Potamogaton Schweinfurthis and *Ceratophyllum demersum* [24]. Overall, the country faces serious problems of resource protection.

Several environmental impacts, at the origin of the deterioration of water quality, result from the invasion of water bodies by vegetation [25]: siltation of hydraulic axes; the formation of caps with loss of hydraulicity; increased evapotranspiration; the threat to adjacent wetlands; the decrease of the dissolved oxygen level. The proliferation of macrophytes can therefore make it difficult to access water, slow down the flow of water in the canals, block the pumps.

Figure 8.
Degradation of water quality in Senegal by invasive plants: (1) Typha australis on the lake of Guiers; (2) freshwater algae in the lake water; (3) invading lake water plan; (4) degradation of the water quality of the lake.

3. Interventions on the water challenges

Despite the water problems facing urban populations, Senegal has sufficient water resources to meet demand if available resources are properly managed [6]. Senegal's internal renewable surface water resources are estimated at 23.8 km^3/year and renewable groundwater resources are in the order of 3.5 km^3/year. The common part between surface water and groundwater is estimated at 1.5 km^3/year and internal renewable water resources estimated at 25.8 km^3/year [26]. The diversity of water resources offers opportunities for exploitation ranging from surface water abstraction to the use of boreholes in areas with limited surface water resources [10].

3.1 Implementation of relevant policies

Like many other countries, Senegal has adopted several policies at the national and regional levels to guide the conservation and management of its water resources. It has put in place crucial reforms in the water sector which have led to the promulgation of certain regulatory texts and conventions, such as Law 81-13 of 4 March 1981 on the Water Code creation of various associations of water resource users [3]. Thus, in 1995, the public authorities give a very marked inflection to the organization of the sector, as well in urban as rural. Indeed, a reform of the urban water subsector was initiated through the Water Sector Project, which led to the separation of drinking water from sanitation. This reform embodied by Law No. 95-10 of 7 April 1995 resulted in the creation of two different entities that are responsible for the management of the sub-sector of urban water: the National Water Company of Senegal (SONES), a heritage company, and Senegalese Waters (SW), operating company, private operator [6, 27].

3.2 Groundwater, an important additional source in urban centers

The exploitation of groundwater through wells/boreholes is widespread but is generally not regulated or monitored in many parts of Africa. As in many cities in Africa, there is a growing demand for groundwater in Senegal's main cities, mainly fueled by boreholes. Senegal has groundwater resources of about 4 billion cubic meters renewable every year and all the drilling currently carried out that pump this resource mobilizes a maximum of 6% of this resource. Thus, in terms of water availability, Senegal is relatively well endowed, especially since this resource is captured at depths that vary around 100 m. Freshwater stored in underground aquifers can be used effectively to divert the consequences of climate change. The availability of groundwater resources and their replenishment rates are uncertain, posing a serious problem for their management and protection [28]. Therefore, in the future, improved regulation and monitoring of groundwater withdrawals, in addition to appropriate management, will be essential for effective and sustainable monitoring of available water resources in Senegal's cities.

For the abstraction of groundwater in Senegal, dewatering works can be grouped into five major systems in order of importance: boreholes and modern motorized wells; modern wells and wells equipped with wind turbines; modern wells and wells equipped with hand pumps; modern wells with manual or animal drainage; traditional wells with manual or animal drainage [29]. Over the past decades, Senegal has therefore made significant efforts on national resources and with the support of its development partners to meet people's drinking water needs from groundwater. However, it must be recognized that despite the large investments, the demand for drinking water is far from being fully covered.

3.3 Monitoring of conditioned water

In the various cities of Senegal, a good number of inhabitants use bottled water in bottles and sachets. From tap water to the bottle, to the plastic bag, there is a ladder of confidence in the quality of the water to drink, while the older practices of water consumption paradoxically provoke an attachment territorial, while presenting a status apart, since they are detached from any commercial thought. This is why it can be said that modern conditioning practices such as bottling and bagging create a new image of drinking water. Calibrated or formatted through models of different capacity, it pays off, and access is through formal and informal commercial distribution networks. Bottled water is present almost everywhere, from the big supermarket sign to the small neighborhood retailer, to the petrol stations, while water in sachets is mostly sold in small shops, in the urban neighborhoods of Dakar (Medina, Rebeuss ...) for example, but also beyond, the capital [30].

Sachet water conditioning in Senegal is mainly in the informal sector of the economy. Anyone who has access to tap water and owns a refrigerator can create a "small business." This is a common practice in working-class neighborhoods. For modest families, it provides extra income. This ranges from bagging water in fine and transparent plastics, without any indication of source or quality of water, to water bags subject to prefectoral authorization with indications of the origin and characteristics of contained water. It is the work of individual and family initiatives, and represents an activity that involves the respect of certain health standards. The conditioning of the water is more a practical necessity, that of providing the body with the occasional need for water, with a taste that is supposed to be better than that of tap water. In Senegal, its consumption reaches significant proportions. In the streets of Dakar, at any time, it is marketed in bottles and especially in small plastic bags, exchanged for parts of 25 or 50 F CFA. Numerous, by the way, are those who have invested in this business, from children to adults, hence the importance of setting up a structure for its supervision.

3.4 Rainwater harvesting

There is a regional imbalance in the recovery and distribution of water, and therefore in water security. In addition, the uneven distribution and variability of rainfall in sub-Saharan Africa impacts the annual water availability of households. In addition, climate change is constantly increasing extreme events such as droughts and floods with disastrous consequences for people's lives. Sub-Saharan Africa has abundant rainfall, but it is not evenly distributed and highly seasonal. Senegal experiences droughts and floods every year. It receives rainfall ranging from 200 mm in the dry parts (Sahelian domain) to more than 1500 mm in the southern and south-eastern parts of the country (southern Sudan) [2]. Senegal's renewable freshwater resources vary considerably with time and region and cannot adequately meet the growing demand in large cities. Water harvesting could be an additional means of alleviating the problems of drought, scarcity and depletion of water resources. Surface water is scarce and groundwater exploitation is often not profitable. As a result, sustainable rainwater harvesting systems can be a very important solution to the problem of water scarcity. However, the collection of rainwater for domestic use in cities is not sustainable due to the configuration of the building and the diversity of activities that pollute the environment. However, for this rainwater to be a solution to the permanent shortages of water currently observed, it should first be captured and then treated before use [11].

3.5 The construction of additional hydraulic infrastructures

Very important results were obtained during the 2005–2015 decade, marked by the implementation of the Millennium Drinking Water and Sanitation Program (PEPAM), both in terms of the definition of policies and strategies, the mobilization of financing and setting up of access to services. However, there are still major obstacles still to be overcome, including the still inefficient management of drinking water quality issues due to localized pollution of certain groundwater or surface water levels, accessibility still average water points due to the low rate of access to water by connection to homes and the average density of distribution points in rural areas ... [7]. To remedy this, the Government of Senegal, under the Ministry of Hydraulics and Sanitation, has launched projects to increase water infrastructure to combat water scarcity. Such infrastructures should make it possible to increase qualitatively and quantitatively access to water and sanitation services, promote sustainable management of water resources, reduce the incidence of water-related diseases, strengthen sector governance through targeted institutional support, with a view to signing a sector budget support program.

4. Strategies to address the water scarcity in Dakar

The shortage of water, temporary or structural, results from a quantitative and/or qualitative insufficiency of the available water resource compared to the demand. His study contributes to the reflection for a better distribution and preservation of water. Water scarcity is a critical issue when it comes to dealing with the sustainable development of societies. A precise study of the different types of conflicts observed is necessary. These occur recurrently between areas of use, commercial and non-market water uses—in practice urban water supply and irrigation [31].

Difficulties in the supply of drinking water in Dakar are also due to a poorly controlled urbanization policy, which translates into a sort of "let do" in the settlement of the populations, which proceed to anarchic constructions, in zones undeveloped. The housing and housing crisis ended up condemning people to a frantic race to find a piece of land [32].

To this problem is added that created by the increase in the number of consumers, which aggravates an already tense situation, due to the natural increase of the population which largely benefits Dakar and the sustained movement of immigration as well as of the rural exodus, the migrants having as their point of departure the "low quarters," already confronted with the precariousness of the system of supply of drinking water. Given the increase in water demand with the high population growth, in a context of climate change, Senegal, like many countries in the world, is beginning to have an imbalance between its water supply and the demand of citizens in perpetual rise. To remedy this, the different actors in charge of water resources issues (government institutions, development partners, civil society and the private sector) must undertake a sustainable management of the water resources that are available on the territory.

For Lacoste [33], "In the third world countries, municipalities in big cities must now respond to many demands for a fairer distribution of water between rich neighborhoods and slums or slums. Some have water in abundance, while the others, where the vast majority of the population lives, have almost none. How, in these conditions, to make reach a maximum of populations with a drinking water? The answer to this question lies in a significant increase in the budgets allocated to the financing of social facilities. Having understood this, the Senegalese authorities in charge of the issue have approached access to drinking water as a public health imperative, which has become a social priority for the country. This is even more true since the links between water and health on the one hand and water and economic development on the other are no longer in doubt [32]. As access to drinking water requires undeniable financial efforts, innovative strategies are being put in place to enable a larger section of the population to benefit. These strategies are based on a concerted approach that brings together non-governmental organizations (NGOs), the private sector and governments in synergy.

Dakar's drinking water supply has long been a major problem in Senegal. Indeed, since the 1980s, Dakar had begun to register a significant deficit in the water supply of its population. From 4% in 1984, the deficit exceeded the threshold of 30% in 1991 to reach the record level of 100,000 m^3/day in 1998. Thus, to prevent this situation from becoming a disaster, Important means have been put in place to manage this deficit, but this cannot completely solve the problem [34]. The schematic flow of the current Dakar water supply system in 2013 is illustrated below in **Table 1** [35].

The history of Dakar's water supply began in 1949, when groundwater from the sub-basaltic aquifers of Mamelles and Point B, as well as the quaternary sand aquifers of Thiaroye, were drilled and developed. Then, in 1960, to meet the increase in demand for water, taking into account the risks of over-pumping aquifers near Dakar, new Paleocene limestone aquifers were exploited at Pout and Sébikhotane. The steady rate of increase in water demand in the following years necessitated the continued development of water resources. Consequently, in 1970, aquifers of the Maastrichtian layer along the northern coastal zone (North Coast) were exploited and the surface waters of the Senegal River were taken from Lake Guiers [35]. In the absence of sufficient local drinking water resources, Dakar is supplied by a water supply system from Guiers Lake. This system transports water from the Senegal River to the capital over 250 km and represents 50% of Dakar's drinking water supply [24]. The strong population growth of recent years has led to the saturation

Water treatment plant		**Year in service**	**Extension**	**Nominal capacity (m^3/d)**	**Hourly volume (m^3/h)**
Ngnith factory		1971	2000 (transition to a theoretical capacity of 60,000 m^3/d)	40,000	1667
Keur Momar Sarr factory		2004	2008 (transition from 65,000 to 95,000 m^3/d) 2011 (transition from 95,000 to 130,000 m^3/d)	130,000	5417
Drilling	Number of drilling	Year in service	Extension	Nominal capacity (m^3/d)	Hourly volume (m^3/h)
Northern littoral drilling (Gueoul at Ndande axis)	9	1999	—	35,000	1591
Kelle/Kebemer drilling	7	from the 1970s	—	30,000	1364
Pout Nord drilling	13	from 1978 (PN6 and PN10)	—	47,248	2148
South Pout drilling	7	from 1979 (PS5)	—	20,000	909
Pout Kirene drilling (including KSW)	4	1993 (PK3, PK5)	—	6000	273
Sebikotane drilling	1	1957	—	4500	205
Thiaroye drilling	2	1951	Not used due to deterioration of water quality	0	0
Point B/Mamelles/Point G drilling	8	1966	—	18,000	818
Booster		Year in service	Extension	Nominal capacity (m^3/d)	Hourly volume (m^3/h)
Mekhe booster		2006	—	233,557	10,155
Carmel booster		2013	—	241,708	10,509
Pumping station		Year in service	Extension	Nominal capacity (m^3/d)	Hourly volume (m^3/h)
Thiaroye factory		1951	—	29,900	1300
Point B factory—Madeleine pumping		1966	—	20,700	900
Point B factory—pumping Mamelles		2006	—	64,400	2800
Point B factory—pumping point		1966		7000	700
Transmission line		Year	Characteristics		

Water treatment plant	Year in service	Extension	Nominal capacity (m^3/d)	Hourly volume (m^3/h)
ALG1 (Ngnith pipe)	1971	DN 1000 PN 25 steel		
ALG2 (driving KMS)	2004	DN 1200 PN 25 cast iron		
800 Sebi	2008	DN 800 cast iron		
600 discharge Thiaroye	1951–1994	DN 600 cast iron		
700 output tanks PTY	1951	DN 700		
800 Mamelles repression	1993	—		
600 South Pout repression	—	DN 600 PN 16 steel		
Tanks	Year	Characteristics		
Tanks of Thies	1971 (R1, R2); 2005 (R3, R4)	25,000 m^3		
Tanks Y-point	1951	10,000 m^3		
Tanks of Madeleines high service	1966	1200 m^3		
Tanks of Madeleines low service	1966	6000 m^3		
Tanks of Mamelles	2003	35,000 m^3		
Tanks of G point	1966	5000 m		

Source: JICA study mission based on information provided by SDE.

Table 1.
Main lines of major structures in the water supply network for the Dakar region.

of production and transfer capacities. Currently, nearly 1 million people in the capital suffer from intermittent service. The Dakar region, which comprises 25% of Senegal's population and concentrates 80% of the country's economic activities, has its water needs estimated at around 320,000 m^3/d, which represents nearly 75% of the total production water supply [35]. In 1993, 80% of water consumption in the Dakar region consisted of groundwater, while the remaining 20% came from Lake Guiers. In 2013, this ratio was reversed due to over-exploitation of groundwater [36].

To ensure an optimal water supply for the city of Dakar and fight against water scarcity, the government has mobilized since 2014 an additional production of 100,000 m^3/day through the realization of 60 boreholes and the rehabilitation of seven others. This additional volume represents 26% of the average daily production (360,000 m^3). From 2014, a peak of 390,000 m^3/day is reached with the commissioning of Bayakh's new drinking water production center in July 2018. This production is provided by the factories of Keur Momar Sarr and Ngnith installed on the site of Guiers Lake (40%) and boreholes of the North Coast, South Pout, Pout Kirène, Kelle-Kébémer and Dakar (60%). With the commissioning of the two Bayakh-Thieudème-Diender and Tassette phases, additional production will reach 179,000 m^3/day overall, or nearly 50% of the capital's peak needs.

From January 2011 to June 2018, production increased by 22%, from 297 million L/day in December 2011 to 355 million L/day in June 2018. The peak of 439 million L/day will be reached in December 2018, i.e., +29% with the commissioning of Bayakh and Tassette, as well as the three new boreholes of Dieuppeul, Yoff and Nord Foire. SONES has implemented the various phases of the Emergency Program with the SDE. Thanks to the impact of this work, the deficit neighborhoods had better access to the drinking water service: Nord Foire, Ouest Foire, CPI, Cité Alternance, Scat Urbam, Grand Yoff, Liberté 6 extension, Mixta, Keur Damel, Socabeg, Cité Léopold Sédar Senghor, Hlm Grand Yoff, part of the Unit 26 of Parcelles Assainies, Toubab Dialaw. In 2017, the Ministry of Hydraulics and Sanitation has developed the Special Program for Drinking Water Supply in Dakar (PSDAK) which is an intermediate solution pending the completion of structural works such as: the third production plant and Keur Momar Sarr drinking water treatment (KMS3) and the des Mamelles seawater desalination plant in Dakar.

The PSDAK has two phases that aim to strengthen production, improve the quality of the water distributed and secure the supply of electricity. The first phase of the PSDAK consists of hydraulic works at Bayakh, covering a battery of five new boreholes, a pumping station, a storage tank of 1500 m^3 and an adduct line of 18.6 km between Bayakh and Rufisque. It has allowed a production of 15,000 m^3/day which is injected into the network and several deficit areas have better access to drinking water in 2018. The second phase of this program consists in particular, of six boreholes, a station of pumping, a reservoir and a large diameter transfer line on the axis Diender-Thieudème. Ultimately, these structures will bring a volume of water of 15,000 m^3/day complementary.

Apart from additional drilling, and social connections provided by SONES, the KMS 3 and the des Mamelles water desalination plant are the keystones of a water security policy. This option will consolidate production and preserve the capital and the Small Coast from any water stress until 2035. These two major projects of the state are committed to the challenge of water, in the perspective of population growth established at 3% annually. The third Keur Momar Sarr plant (KMS 3) should cover the drinking water needs of the Dakar populations, the new urban center of Diamniadio, the Rose Lake, as well as all the localities crossed by the Lac de Guiers pipeline from 2021 It is expected to represent in 2020 more than 20% of the drinking water supply capacity of the water supply system from Guiers Lake. As for the other structuring project that is the Mamelles seawater desalination plant, with a capacity of 50,000 m^3/day expandable to 100,000 m^3, the water problems of Dakar will be conjugated to the past.

In order to ensure the supply of water to cities, it is therefore essential to improve the availability of sustainable water supply, the conservation and restoration of water bodies as well as strategic investments in additional water infrastructures. Additional water facilities would help increase water storage capacity for long-term uses and avoid recurring disasters such as scarcity. In addition, urban water and sanitation companies should prioritize the construction of efficient wastewater treatment plants to facilitate the treatment and reuse of water.

5. Conclusions

In the face of strong urban growth in Africa as a whole, people's water supply is often lagging behind. Many African cities find it difficult to provide adequate water services to the growing number of occupants. The demand for clean and adequate water is increasing due to population growth and the global obligation to achieve the Sustainable Development Goals, including Goal 6: "Ensuring access for all to water and sanitation services managed sanitation" [37].

Senegal suffers from a chronic water crisis due to various causes including drought, landscape degradation, floods, contamination and unprecedented population growth. If solutions exist against mismanagement and water pollution, the main problem lies in the frequency and severity of extreme events such as droughts and floods due to ongoing climate change, phenomena that will likely be more unpredictable in the future. Adequate provision of drinking water to populations could also be strongly influenced by environmental pollution. Therefore, to preserve water security, it is necessary to focus on the protection of sources, the more judicious use of fertilizers and pesticides, the reduction of domestic and industrial pollution as fundamental elements of the complete water management strategy.

In order to achieve the Sustainable Development Goals (SDG 6 in particular), new strategies for the sustainable management of water resources are needed. In fact, in urban areas that use large quantities of water (which also puts a lot of pressure on the country's resources), the modernization of water infrastructure is an important step to implement for the sustainable preservation of water. Pure water. These strategies must also take into account the improvement of access and access to drinking water, the fight against waste of the resource, the treatment and reuse of water used for agricultural purposes, the storage of water in period of rainfall abundance and its reuse in times of scarcity, preservation of aquatic ecosystems.

On the issue of the recurring water shortage in Dakar, it is recommended a device with the following objectives: to reinforce the hydraulic equipment; reduce the vulnerability of people and goods; appreciate, treat and reduce the risk of water scarcity; put in place the required prevention, response and recovery measures; maintain essential activities and services; identify external actors and integrate them into the planning process.

Acknowledgements

We thank all the national and international organizations that provide us with the data that helped write this article.

Conflict of interest

There are no conflicts of interest for this document.

Notes/thanks/other declarations

There are no other notes/thanks/statements.

Author details

Cheikh Faye
Department of Geography, U.F.R. Science and Technology, UASZ, Laboratory of Geomatics and Environment, Ziguinchor, Sénégal

*Address all correspondence to: cheikh.faye@univ-zig.sn

References

[1] Agence Nationale de la Statistique et de la Démographie. Recensement Général de la Population et de l'Habitat, de l'Agriculture et de l'Elevage (RGPHAE); Rapport provisoire, 2013; mars 2014. 36 p

[2] Faye C, Ndiaye A. et Mbaye I. Une évaluation comparative des séquences de sècheresse météorologique par indices, par échelles de temps et par domaines climatiques au Sénégal. Journal. wat. env. sci. 2017;**1**(1):11-28

[3] Faye C, Dieye S. Valorization of water resources in Senegal for economic, social and sustainable development. African Journal of Environmental Science and Technology. 2018;**12**(12):449-460

[4] Sane M. Note sur les ressources en eaux du Sénégal: zones potentielles pour le transfert d'eau. Directeur de l'hydraulique, République du Sénégal; 2015. 8 p

[5] pS-Eau. Réseau Sénégal, Les actualités du secteur de l'eau et de l'assainissement au Sénégal. Newsletter; 2015. 9 p

[6] CONGAD. Livre bleu "L'eau, la vie, le développement humain" Rapport pays. Sénégal: Document de consultation fourni par SENAGROSOL CONSULT; 2009. 72 p

[7] Ministère de l'Hydraulique et de l'Assainissement du Sénégal (MHAS). Parvenir aux ODD pour l'eau, l'assainissement et l'hygiène Aperçu de la situation au Sénégal: eau, assainissement et hygiene; 2015. 8 p

[8] UN World Water Development Report. 2nd UN World Water Development Report. Citation Time(s); 2006. p. 4

[9] Ministère de l'Hydraulique/ DGPRE. Plan d'Action de Gestion intégrée des ressources en eau du Sénégal: Global Water Partnership West Africa; 2007. 61 p

[10] Faye C, Gomis EN, Dieye S. Current situation and sustainable development of water resources in Senegal. Ecological Engineering and Environment Protection. 2019;**1**:5-16

[11] Ondigo DA, Kavoo AM, Kebwaro J. Water resources and management under increasing urban demography: A Kenyan perspective—A review. Journal of Water Resource and Protection. 2018;**10**:919-938

[12] Agence Nationale de la Statistique et de la Démographie. Situation économique et sociale du Sénégal en 2015; 2013. 34 p

[13] Diallo A. Journée d'information et d'échanges sur le secteur de l'eau et de l'assainissement au Sénégal: Ministère de l'Hydraulique et de l'Assainissement, 20 juin 2016. Paris, France: Pavillon de l'Eau; 2016. 49 p

[14] World Bank Database. 2018. Available from: https://data.worldbank.org/indicator/SP.POP.TOTL?locations=SN&view=chart

[15] WWF Report 2011, Big cities. Big water. Big challenges. Water in an urbanizing world. [Citation Time]

[16] UNRISD. United Nations Research Institute for Social Development, Environmental Degradation and Social Integration: Paper No. 3, World Summit for Social Development (November, 1994); 1994. 23 p

[17] Faye C. Les défis de la pollution de l'eau, une menace pour la sante publique: atouts et défauts des lois et politiques de l'eau au Sénégal. Larhyss Journal. 2017;**32**:107-126

[18] Mayers J, Batchelor C, Bond I, Hope RA, Morrison E, Wheeler B. Water Ecosystem Services and Poverty under Climate Change: Key Issues and Research Priorities. Natural Resource Issues No. 17. London: International Institute for Environment and Development; 2009. 86 p

[19] UNESCO. Water for People, Water for Life: World Water Assessment Program; 2003. 36 p

[20] OMVS. Projet FEM/Bassin du fleuve Sénégal, 2008. Plan d'action stratégique de gestion des problèmes environnementaux prioritaires du bassin du fleuve Sénégal: Version finale; 2008. 133 p

[21] WWAP. "Senegal river case study": in Premier rapport du Programme Mondial pour l'évaluation des ressources en eau (WWAP); 2003

[22] Sène AM. Développement durable et impacts des politiques publiques de gestion de la vallee du fleuve Sénégal: Du régional au local », VertigO—la revue électronique en sciences de l'environnement [Online]. 2009;**9**(3). Disponible sur: http://vertigo.revues.org/9221

[23] Awaïss A. L'Initiative Bassin du Niger (IBN): développement durable et gestion intégrée d'un grand fleuve. Afrique contemporaine. 2003;**2**(206):179-203

[24] Faye C. le lac de Guiers: Etude du régime et des bilans hydrologique et hydrochimique: quelles incidences écologiques et socio-économiques dans la zone: Mémoire de maitrise de géographie UCAD; 2007. 155 p

[25] Ministère de l'Environnement et du Développement Durable. Situation de référence de la biodiversité dans les zones de peuplement en Typha: PNEEB/TYPHA; 2014. 47 p

[26] FAO. Site web AQUASTAT. Organisation des Nations Unies pour l'alimentation et l'agriculture. Site consulté le 21/05/2017.2016. Disponible sur: http://www.fao.org/statistics/fr/

[27] Tine J. Etat des lieux des collectifs, plateformes et réseaux d'organisation de la société civile du secteur eau et assainissement dans 7 pays de l'Afrique de l'ouest et du centre: Rapport Sénégal; 2009. 55 p

[28] Van der GJ. Groundwater and Global Change: Trends, Opportunities and Challenges. Paris: UNESCO Publishing; 2012. [Citation Time(s): 1]; 44 p

[29] République du Sénégal. Etude de faisabilité des forages manuels, identification des zones potentiellement favorable: PRATICA, UNICEF; non daté. 45 p

[30] Valentin M. Bouteilles et sachets en plastique. Pratiques et impacts des modes de consommation d'eau à boire au Sénégal. Autrepart. 2010;**55**:57-70

[31] Honegger AR, Bravard J-P. La pénurie d'eau, donnée naturelle ou question sociale? Géocarrefour. 2005;**80**(4):257-260

[32] Bignoumba GS. L'appro visionnement en eau à Libreville et Dakar. In: Villes en parallèle, n°40-41, Janvier 2007. Villes du Gabon; 2007. pp. 180-195

[33] Lacoste Y. L'eau dans le monde: les batailles pour la vie: Coll. Petite Encyclopédie Larousse, Baume-les-Dames; 2003. 127 p

[34] Faye MD. Etude de Sectorisation d'un réseau Urbain d'eau Potable de Dakar Ville: Cas de Fann Hock: Mémoire de Master; 21E; 2012. 84 p

[35] PDU Dakar. Plan Directeur d'Urbanisme de Dakar et Ses Environs

Horizon 2035; Rapport Final: Volume I; 2014. 103 p

[36] AFD. Programme d'urgence pour la sécurisation de l'alimentation en eau potable de Dakar. Note de communication publique d'opération république du Sénégal: CSN1446; 2014. 11 p

[37] Ndour N. Etat d'avancement du processus de suivi de l'ODD6 de l'initiative GEMI au Sénégal. Conférence internationale sur les Objectifs de Développement Durable (ODD): Quel agenda pour le Sénégal ?12-13 octobre 2016, Dakar, Sénégal; 2016. 14 p

Section 3

Evaluation and Treatment

Chapter 9

GIS and Statistical Evaluation of Fluoride Content in Southern Part of Upper Rasyan Aquifer, Taiz, Yemen

Ramzy Saeed Mahbob Naser, Mohammed El Bakkali and Driss Belghyti

Abstract

Fluorosis continues to be an endemic problem in Yemen. More areas are being affected by fluorosis in different parts of this country. The present study aims to identify the intensity and the spatial extent of fluoride concentration in groundwater of the southern part of the upper Wadi Rasyan, Taiz, Yemen. 93 sampling points were selected; the sampling included all types of sources of groundwater and all types of aquifers. The results show that 71% of samples exceed the WHO drinking water guidelines value of 1.5 mg/l, and there are wide variation for groundwater's content of fluoride in the same aquifer (whether, volcanic and alluvial) and in the same of groundwater type, and these variations between the different water types or between the different depths of water (alluvial and volcanic aquifers) are not significantly different. The high concentration of fluoride in groundwater of the volcanic aquifer is likely because of the nature of geology formations by the water-rock interaction result of long-time residence of water in contact with the geology formation. The high concentration of fluoride in the alluvial aquifer likely resulting the waste of urban and industrial activates sources, the over exploration of groundwater, the arid climatic and the activities agriculture.

Keywords: fluoride, spatial distribution, GIS, Al-Burayhi and Hedran, Al-Hawban, upper Wadi Rasyan, Taiz and Yemen

1. Introduction

Fluorine has the highest chemical reactivity among all known elements and occurs mainly as free fluoride ions in natural waters, although some fluoride complexes also exist under specific conditions [1]. In groundwater, the natural concentration of fluoride depends on the geological, chemical and physical characteristics of the aquifer, the porosity and acidity of the soil and rocks, the temperature and the action of other chemical elements [2]. Fluoride ion in drinking water is known for both beneficial and detrimental effects on health. Fluoride in small amounts is an essential component for normal mineralization of bones and formation of dental enamel [3]. However, excessive intake of fluoride can cause dental and skeleton fluorosis [4, 5]. Due to its strong electronegativity, fluoride is attracted by positively charged calcium in teeth and bones [6]. Fluorosis is a considerable health problem worldwide, which is afflicting millions of people in many areas of the world, for example, East Africa [7–9] and India [10–12]. According to World Health Organization (WHO) Guidelines for Drinking Water Quality [13], the limit value for fluoride is 1.5 mg/L. The value of 1.5 mg/L is a guiding value, which may be changed based on climatic conditions like temperature, humidity, volume of water intake, fluoride from other sources, etc. for different regions of the world [14]. The source of water supplies in Yemen is mostly from groundwater accumulated during previous and current times [15]. Fluorosis continues to be an endemic problem in Yemen. More areas are being affected by fluorosis in different parts of the country. Recently, a report from General Authority of Rural Water Projects (GARWP) indicates markedly increasing in fluoride content in groundwater (between 2000 and 2006) in districts of some governorates such as Sana'a, Ibb, Dhamar, Taiz, Al-Dhalei and Raimah. The highest fluoride concentration in drinking water was reported in some districts of Sana'a governorate, especially Sanhan [16]. Most Yemenis dwelling in rural areas use deep well water for drinking and household works, and a large number of these wells are contaminated with fluoride in a concentration of 2.5–32 mg [14]. The present study aims to identify the intensity and the spatial extent of the existing groundwater contamination by fluoride in the study area and tries to identify sources pollution responsible for the current pollution of the affected areas through an analytical study in the southern part of the upper Wadi Rasyan of Taiz governorate in Yemen.

2. Materials and methods

To achieve the objectives mentioned above, there has been:

1. Identifying and understanding of the characteristics of the study area (topographic and hydrological analysis): location, topography and hydrological characteristics using arc Map GIS.

2. Inventing sources of pollution and production of their maps: inventory of number, type and intensity of human activities and the village's distribution that is likely to contaminate the groundwater in the study area, view inventory results on the map using arc Map GIS and using this map in the interpretation of the results of the spatial assessment of groundwater quality of the study area.

3. Inventing of wells in the study area and displaying them on the map using arc Map GIS.

4. Determining sampling points based on type of wells (dug well and bore well), type of aquifers (alluvial and volcanic), the different depths (from 9 to 500 m) and their location according to the hydrology system and the pollution sources in order to appropriate selection of sampling point and production of the map of sampling points, by using the arc Map GIS.

5. Taking, transporting and analyzing samples.

6. Data processing and interpreting by using arc Map GIS and Minitab 18 program software.

7. Viewing the results of the analysis on the maps in order to know the spatial distribution of fluoride concentration in groundwater of the study area. The spatial distribution of fluoride in groundwater samples in the study area is represented as a thematic layer using IDW tool in the arc Map GIS software program that was used to the prediction of an unknown value for fluoride of the rest of the study area that was not covered by analysis and thus gave the spatial distribution of the fluoride that used in assessing the suitability of groundwater for drinking in the study area as a whole.

8. Using the results of the groundwater assessment quality to propose alternative strategies to deal with groundwater.

9. Preparing the final reports (article).

2.1 Sampling

The sampling was collected in polyethylene bottles of 1000-ml capacity after rinsed with distilled water and the water of the well, through months in August, September and October, 2014. The fluoride concentration of groundwater samples was determined using DR 2800 spectrophotometer.

2.2 Statistical methods

The Fisher test was used when comparing dichotomous data separately and Pearson's correlation coefficient for continuous variables. On the other hand, after verifying the hypotheses of normality and homogeneity, we used the nonparametric Kruskal-Wallis H to test whether three or more samples were drawn from the same population, or from populations with identical characteristics (distribution with the same median). An analysis of variance was used to study the difference in means between the different samples greater than or equal to three and in the multivariate analysis between our samples two by two we chose the Bonferroni test. In the study, Al-Hawban, Al-Burayhi and Hedran and Al-Dhabab sub-basins were all different samples. After, we performed logistic regression analysis. Fluoride was included as a dichotomous variable (lower or greater than 1.5 mg/L). Other variables with p-values < 0.2 in the univariate analysis were entered into the multivariate logistic regression model, which where taken into account in the multivariate logistic analysis. We studied the cause-effect association between the fluoride (lower or greater than 1.5 g/ml) and the included variables using odds ratio (OR). In our first model (crude model), we were satisfied only on the univariate analysis between each variable (independent factors) and the dichotomous concentration of fluoride (dependent variable). In a second model, we performed a simultaneous analysis between the independent variables and the dichotomous dependent variable of fluoride. In order to assess the accuracy of the estimates, we have indicated the 95% confidence interval (IC to 95%) of the average data. A p-value of less than 0.05 at 95% confidence level was considered as statistically significant.

3. Results

3.1 Results characteristics analysis of the study area

3.1.1 Location of the study area

The study area represents the southern part of the Upper Wadi Rasyan catchment area, Taiz governorate, Yemen. The study area is estimated at 472 square kilometers which is densely populated and includes Taiz city which represented the third largest and important cities in Yemen (**Figure 1**).

3.1.2 Results of topography analysis of the study area

The results of the topographic analysis (morphology, elevation and slopes degree) of the study area are illustrated in **Figures 2–4**.

From the topographic analysis of the study area, we find that the group of mountains and plateaus that surrounding the study area made it semi-closed. Therefore,

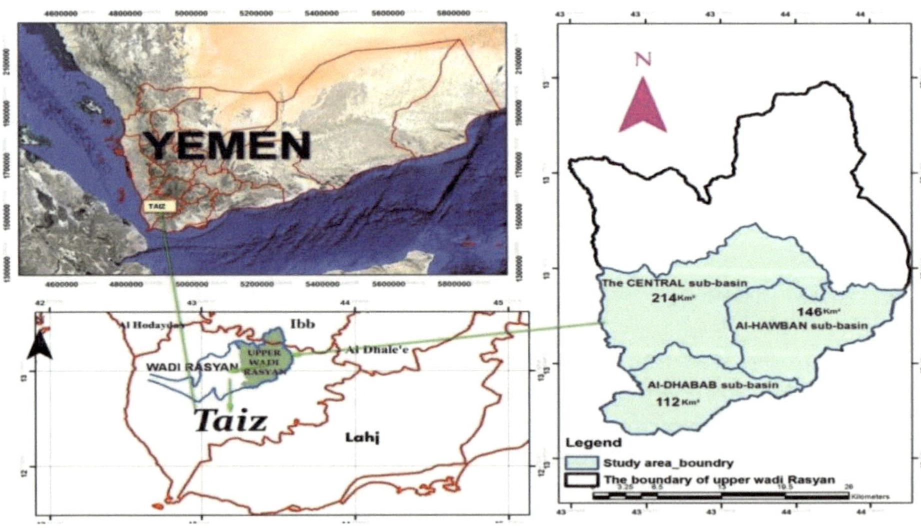

Figure 1.
Location of the study area [17].

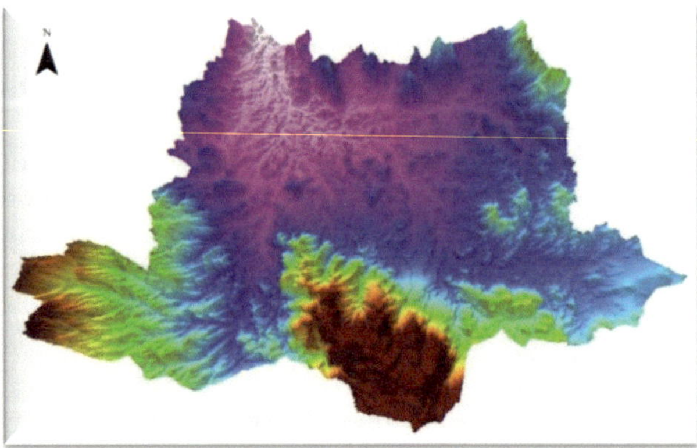

Figure 2.
Topography of the study area.

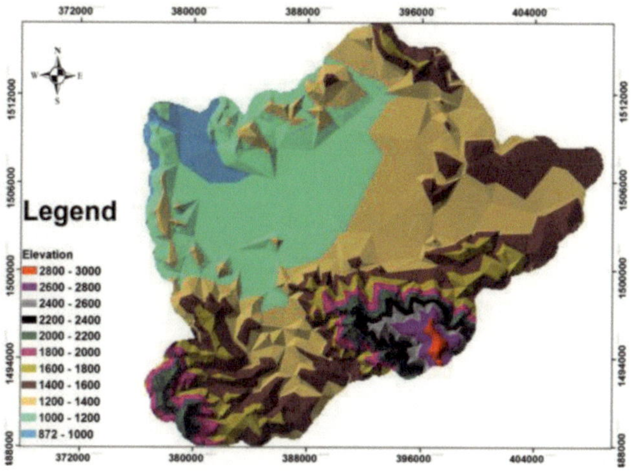

Figure 3.
The elevation in the study area.

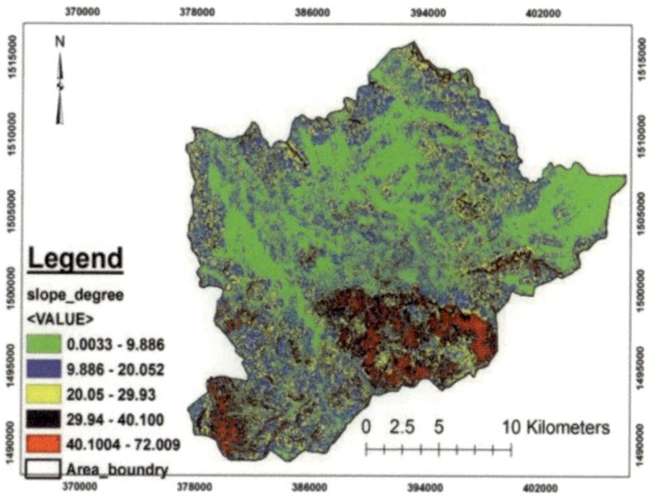

Figure 4.
The slope degree in the study area.

the north-west corner of the study area formed the outlet for the runoff network as shown in **Figure 5**. According to the digital elevation model of the study area that was obtained from the NASA site and topographic analysis, the elevations in the study area concentrated in the south and southwest, where the height of the mountain of Sabir was up to 3000 m above sea level, whereas Jabal Habashi has a maximum height of 2400 m above sea level. The lowest elevation is located in the north-west and is 872 m above sea level, and in the east, elevations are between 1200 and 1600 m above sea level (**Figure 3**). **Figure 4** shows the map of the degrees of slope in the study area. The degree of slope determines the flow intensity of the floods and, therefore, the extent of water (or water and pollutants) infiltration into the ground.

3.1.3 Results of hydrology analysis of the study area

The results of the hydrological analysis [the hydrological limits, direction of surface runoff and then the flow direction of pollutant at the surface (hydrology system) and the hydrological level (main valleys and its tributaries)] of the study

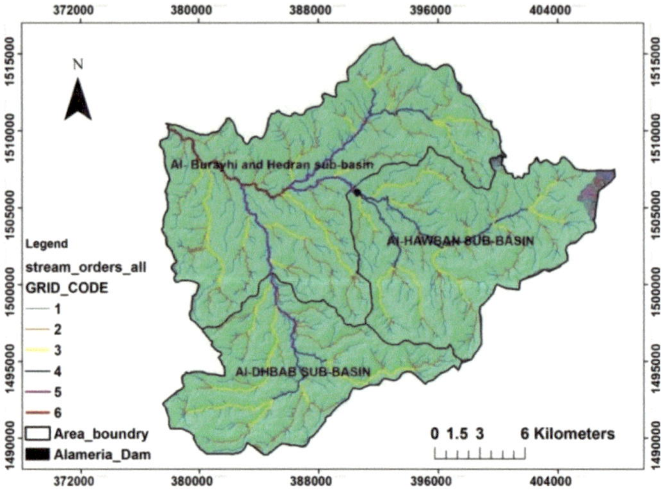

Figure 5.
Hydrology system in the study area.

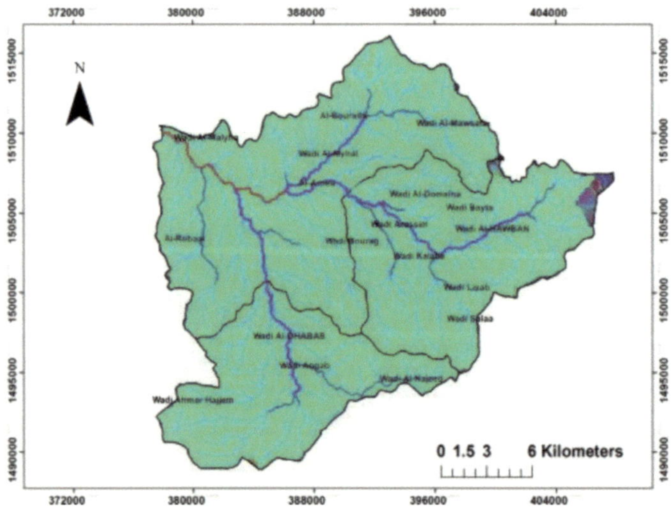

Figure 6.
Main valleys and its tributaries in the study area.

area are illustrated in **Figures 5** and **6**. From the hydrology map of the study area, we found that the watershed drainage (the hydrology system) in the study area is dendritic and the direction of water surface is into the north-west corner of the study area. Based on hydrological characteristics, the study area has been divided into three main sub-catchment area or sub-basins as shown below:

1. Al-Hawban sub-basin has 146 km^2.

2. Al-Dhabab sub-basin has 112 km^2.

3. Al-Burayhi and Hedran sub-basin (Central sub-basin) has 214 km^2.

3.1.4 Geomorphology and geology of the study area

The geology of Yemen is a part of the Arab Shield which consists of a Proterozoic crystalline subsoil covered by a Paleogene sedimentary sequence (Mesozoic sediments). In the west of the country, the sedimentary sequence is covered by Yemen volcanic (Cenozoic volcanic) [18]. The study area occupies the southwestern corner of the Arabian Shield, and the geological complexity of the region is mainly due to its position at the junction of the Red Sea and the Gulf of Aden rift systems. Geological map and geological cross sections for the study area were derived from the geological map of the upper Wadi Rasyan, scale 1: 100,000; that prepared by Dar Yemen Consulting Company [19]. As shown in **Figure 7**, the geomorphology of the study area is dominated by the tertiary volcano that covers most of the region.

Geomorphology of the study area was characterized as modern rock units, which was formed during the Cenozoic (Cenozoic volcanic group), which was formed by a series of eruptions and volcanic eruptions that were affected by Yemen in general and the province of Taiz, especially during the third geological age as a result of movements of the successive continental shelf separation tectonics along the fault line of large expanses of the Gulf of Aden, in the southern Dead Sea to north and the emergence of the Red Sea gorge and the separation of the Arabian plate from the African plate where it was accompanied by the emergence of streaks parallel to the axis of the Red Sea, which represented the levels of weakness and

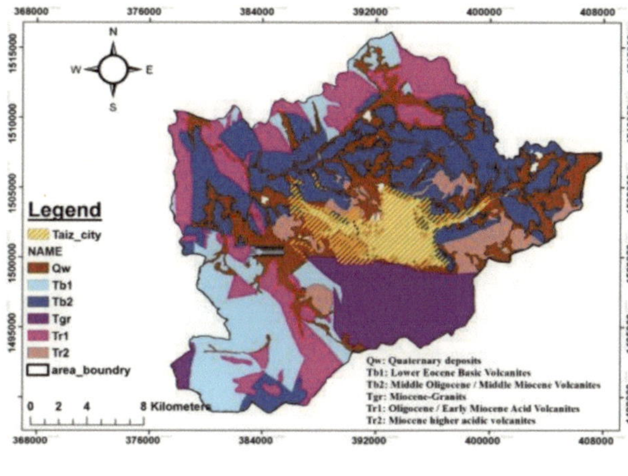

Figure 7.
Geomorphology of study area (the source of basic map [19]).

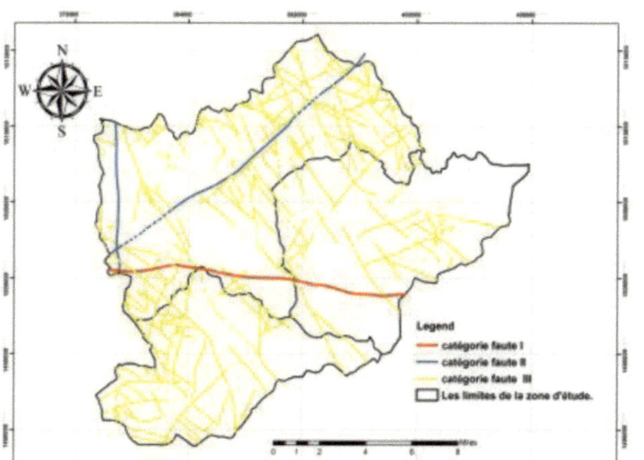

Figure 8.
Geology faults in the study area (the source of basic map [19]).

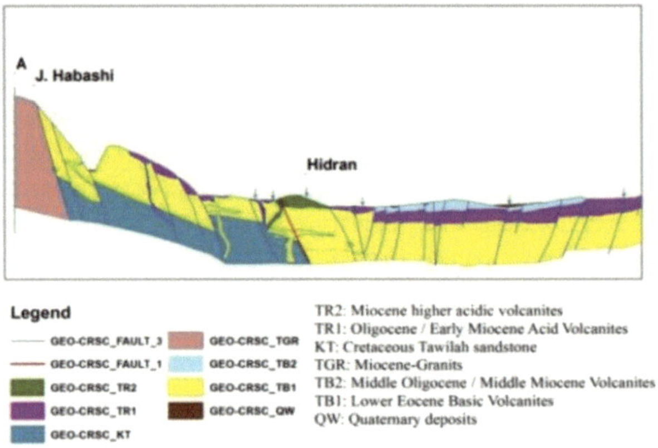

Figure 9.
The geology cross-section of Jabal Habashi to Hedran (the source of basic map [19]).

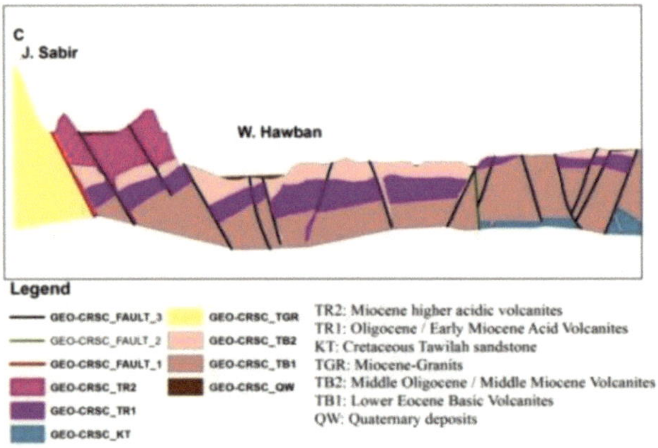

Figure 10.
The geology cross-section of Jabal Sabir to Al-Hawban (the source of basic map [19]).

pathways of magmatic systems [20]. A major crack extends from the east to the west of the study area. There are also local cracks stretching group north-west to south-east perpendicular to the main fault (**Figure 8**).

According to [19], the geology of the study area consists of Cretaceous Tawilah Sandstone (Kt), Lower Basalt (Tb1), Low Volcanic Acids (Tr1), Basic Volcanic Medium (Tb2), Second Volcanic Acids (Tr2), Granite (Tgr): *granite in the mountain of Sabir contains some older volcanic rocks* and finally Quaternary (QW): Wadi sediments are deposited by seasonal floods and wind-deposited soils derived from the alteration of volcanic ash and tufa mainly of (Tr1). Thickness can be up to 70 m. **Figures 9** and **10** illustrate geological cross-sections of some study areas.

3.1.5 Hydrogeology of the study area

According to [19], groundwater in the study area is being produced from three aquifers: the Quaternary alluvium, the Tertiary fractured volcanic and the Cretaceous Tawilah Sandstone. Cretaceous Tawilah Sandstone in the study area is located in the lower classes, in the southwest of the study area and extends to the north, as shown in **Figure 11**.

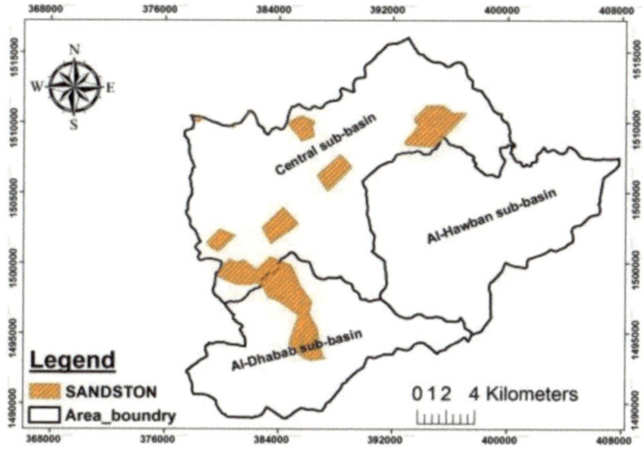

Figure 11.
Cretaceous Tawilah sandstone in the study area (the source of basic map [19]).

Alluvial aquifers form the highest and shallow aquifers in the region. These sediments exist along the Wadis path (**Figure 6**). The total thickness rarely exceeds 30–40 m, but they can locally reach considerable thicknesses (up to 70 m). Hydraulic properties vary from site to other. Intergranular groundwater flow is dominant [19].

Volcanic aquifers consist of the tertiary volcanic sequence. The thickness of this sequence may exceed 600–700 m. Groundwater flows mainly in this type of aquifers through the cracks/faults. The sandstone aquifer includes Tawilah Sandstone, Southwest of the study area (Al-Dhabab), and this formation is soaked to an expected depth of more than 500 m. In general, sandstone is largely silicified and fractured, so that the dominant groundwater flow in this aquifer is of intergranular type and mixed fracture. The quality of this aquifer is excellent to good [19].

3.1.5.1 Groundwater recharge

Groundwater aquifers in the study area are recharged by many sources of water, as follows:

1. In Al-Hawban sub-basin, the aquifers are recharged with rainwater (either pure or loaded with wastewater that is disposed of in floodwaters), wastewater disposed of in sewers, industrial wastewater in the eastern part of the sub-basin where there is an industrial food complex.

2. In Al-Burayhi and Hedran sub-basin, the aquifers are recharge with the floods coming from the Al-Hawban sub-basin that loaded with liquid and solid waste types, domestic wastewater that is transported from the city of Taiz across the sewerage network is deposited in sedimentation ponds in this sub-basin, wastewater used to irrigate crops in this sub-basin, irrigation water, which goes downloaded with high concentrations of salts, sedimentation ponds for industrial wastewater in the western part of the sub-basin, the floods that coming from the Al-Dhabab sub-basin to the south-west of the Al-Burayhi and Hedran basin, these floods reach the Al-Burayhi and Hedran sub-basin, which is mostly pure but soon to be contaminated by the remnants of industrial activities located west of the Al-Burayhi and Hedran sub-basin.

3. In the Al-Dhabab sub-basin, rainwater is almost the only source of recharge for the aquifers.

3.1.5.2 Groundwater flow

According to [19], in general, the groundwater movement in the study area is a function of the hydrological system into the north-west corner of the study area (toward the Al-Burayhi and Hedran sub-basin). In the volcanic aquifers, the direction of the groundwater movement is subject to the direction of the faults.

3.1.6 Results of inventory's contamination source in the study area

The results and spatial distribution maps of village and inventory's contamination source in the study area are shown in **Figures 12** and **13**.

The sampling sites are illustrated in **Figure 14**.

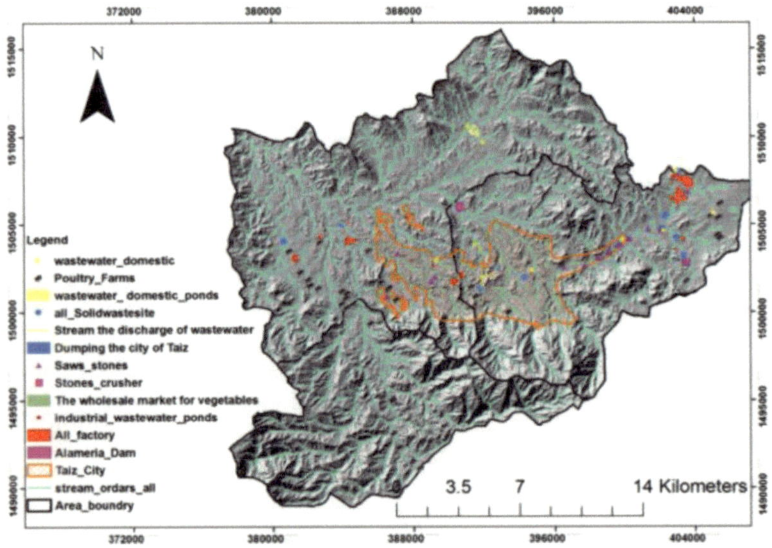

Figure 12.
Spatial distribution of pollution sources in the study area.

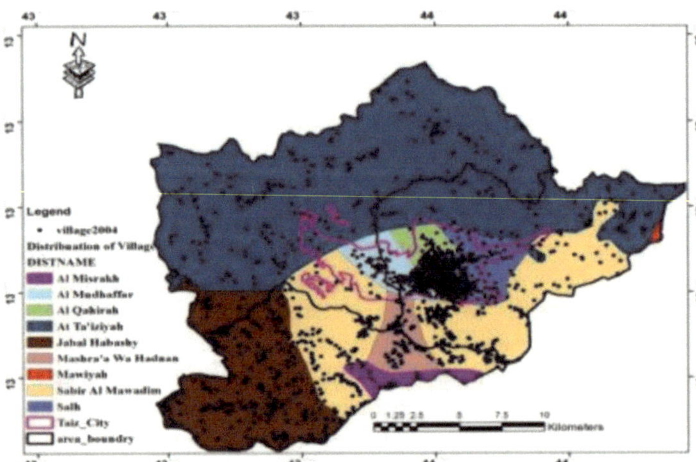

Figure 13.
Village's distribution in the study area.

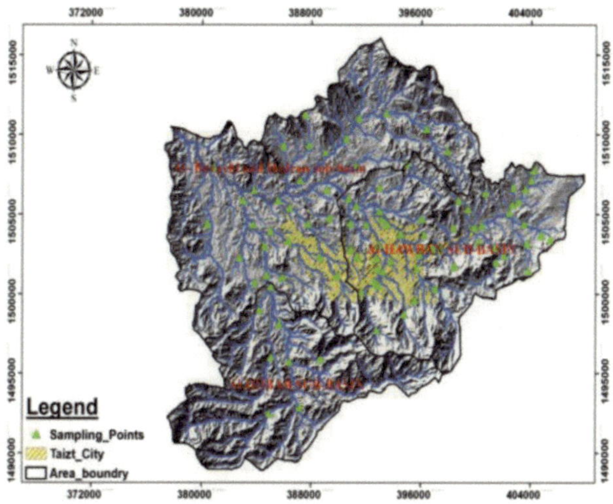

Figure 14.
Sampling sites.

3.2 Results analysis of fluoride's concentration in groundwater of the study area

3.2.1 Descriptive statistic of results

A summarized statistic descriptive of results of fluoride's concentration (mg/L) in groundwater of the study area is shown in **Table 1**, and the minimum, maximum, means and standard deviation of results based on each sub-basin are illustrated in **Table 2**.

We used boxplot tool in order to provide a simplified presentation of how the values of fluoride's concentration are distributed, the boxplot (**Figure 15**) illustrated, the values' distributions are dissimilarities in their distribution in three sub-basins.

Variable	Min	Q1	Median	Q3	Max	Mean	StDev
F- mg/L	0.100	1.470	1.890	2.980	6.000	2.353	1.449

Table 1.
Descriptive statistics of F- mg/L for all samples.

	Al-Hawban, sub-basin	Al-Burayhi and Hedran, sub-basin	Al-Dhabab, sub-basin
No. of wells	56	29	8
Minimum	0.98	0.1	0.58
Maximum	5.81	5.45	1.11
Mean	2.32	2.66	0.85
S.D	1.15	1.26	0.189

SD, standard deviation.

Table 2.
Descriptive statistics of fluoride values mg/L in groundwater samples of the study area by sub-basin.

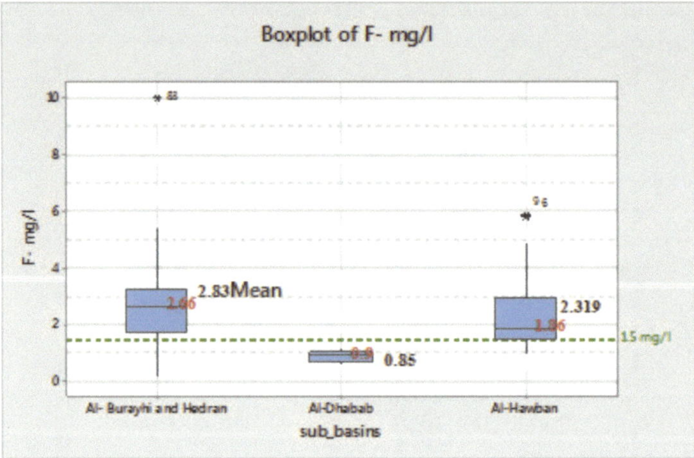

Figure 15.
Box plot to provide a simplified presentation of how the values of fluoride's concentration are distributed in groundwater of three sub-basins.

3.2.2 The correlation analysis

The correlation analysis between the fluoride concentration and the different physicochemical parameters showed that the fluoride concentration is positively correlated with the Cl, EC, TDS, K, Na, Mg and T. H at the significance level of 0.01 and with the parameters Ca and HCO3 at the significance level of 0.05 (**Table 3**).

3.2.3 The relationship between the concentration of fluoride and the water type

Figure 16 illustrated how the values of fluoride's concentration are distributed according to water type, and **Figure 17** show the comparison of the mean of F-mg/L according to water type with 95% confidence interval.

From the results of fluoride contain in the 93 samples, we found that the very high fluoride's concentration (4.501–6 mg/L) was associated with the water type both of Na-Cl (5 samples), Na-HCO3 (1 sample) and Mg-Cl (1 sample), the high fluoride's concentration (3.01–4.5 mg/L) was associated with the water type both of Na-Cl (8), Na-SO4 (3), Mg-Cl (1), Na-HCO3 (2) and Ca-HCO3 (1), the moderately abundant of fluoride's concentrations (1.5–3.0 mg/L) was associated with the water type both of Ca-Cl (4), Ca-HCO3 (2), Ca-SO4 (2), Mg-Cl (2), Mg-SO4 (1), Na-Cl (13), Na-HCO3 (16) and Na-SO4 (4), the optimal fluoride's concentration (0.5–1.5 mg/L) was associated with the water type both of Ca-HCO3 (5), Mg-HCO3 (5), Na-Cl (8), Na-HCO3 (7) and Na-SO4 (1) and one sample with water type of Na-SO4 has fluoride's concentration lower than 0.5 mg/L as shown in **Table 4** and **Figure 19**. The distribution of water type according to sub-basins of the study area is illustrated in **Figure 18**, and the spatial distribution of both of water type and fluoride concentrations is illustrated in **Figure 19**.

There are variations for the fluoride concentration in the same of water type as shown in **Table 4**. Normal concentration of fluoride in Al-Dhabab sub-basin was associated with Ca-HCO3 and Mg-HCO3 water type; however, the sources of samples (wells No. 32 and 43, and spring No. 44) in Al-Hawban with the Ca-HCO3 water type showed high values of fluoride concentration 3.66, 2.36 and 1.96 mg/L containing fluoride concentration that exceeds the WHO drinking water, respectively. These sources are located in the southwest of Al-Hawban sub-basin, likely the sources of fluoride results of geology formation of the Sabir's mountain granits,

		C°	pH	EC (µs/cm)	TDS (mg/L)	T.H (mg/L)	T.A (mg/L)	Ca (mg/L)	Mg (mg/L)	Na (mg/L)	K (mg/L)	Fe (mg/L)	Cl (mg/L)	SO4 (mg/L)	HCO3 (mg/L)	NO3 (mg/L)	F- (mg/L)
F- mg/L	P C	0.086	−0.055	0.491**	0.491**	0.394**	0.014	0.211*	0.427**	0.453**	0.489**	0.072	0.536**	0.168	0.262*	0.014	1
	Sig. (2-tailed)	0.414	0.602	0.000	0.000	0.000	0.897	0.043	0.000	0.000	0.000	0.490	0.000	0.107	0.012	0.895	
	No	93	93	93	93	92	93	93	93	93	93	93	93	93	92	93	93

PC, Pearson Correlation.
* *Correlation is significant at the 0.05 level (two-tailed).*
** *Correlation is significant at the 0.01 level (two-tailed).*

Table 3.
Correlation between fluoride and different physico-chemical parameters.

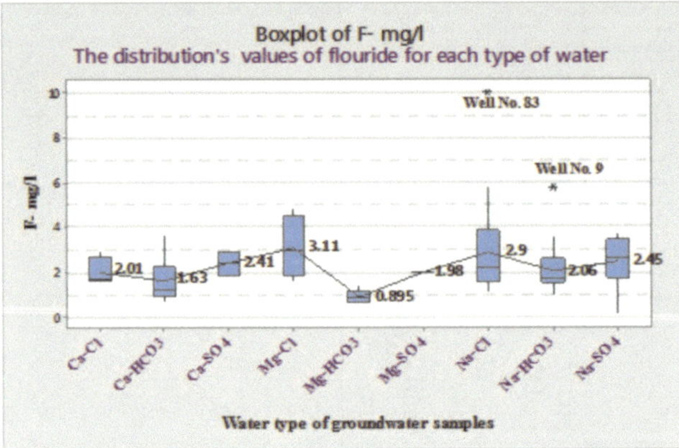

Figure 16.
The distribution's values of fluoride according to water type.

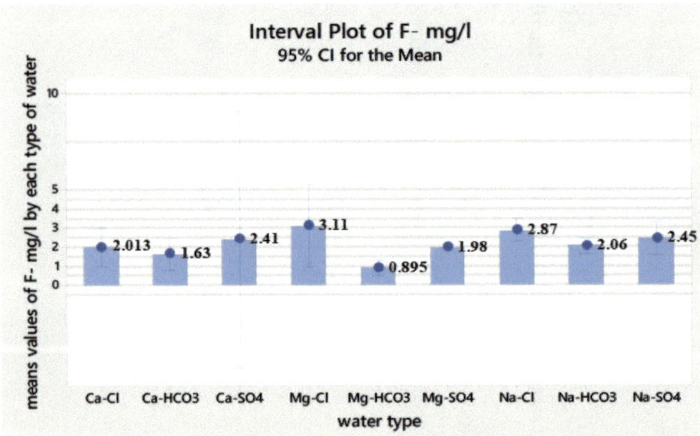

Figure 17.
Comparison of the mean of F- mg/L according to water type with 95% confidence interval.

Concentration of fluoride (mg/L)	Type of water and number of samples
0.5–1.5	Ca-HCO3 (5), Mg-HCO3 (5), Na-Cl (8), Na-HCO3 (7) and Na-SO4 (2)
1.5–3.0	Ca-Cl (4), Ca-HCO3 (2), Ca-SO4 (2), Mg-Cl (2), Mg-SO4 (1), Na-Cl (13), Na-HCO3 (16) and Na-SO4 (4)
3.01–4.5	Na-Cl (8), Na-SO4 (3), Mg-Cl (1), Na-HCO3 (2) and Ca-HCO3 (1)
4.501–10.0	Na-Cl (5), Na-HCO3 (1) and Mg-Cl (1)

Table 4.
Classification of F- mg/L and their relation to the water type.

according to [21]. The groundwater with high concentration of fluoride is associated with the granits rocks.

For the Na-Cl water type, we found that 26 samples out of 34 samples have high concentration of fluoride, 12 samples located in the Al-Burayhi and Hedran sub-basin [(wells No. 83, 80, 78, 77, 68, 65, 88, 82, 87, 79, 75 and 92), where 4 samples of them (wells No. 83, 80, 78 and 77) have 4 of highest values of fluoride's concentration of out of 5 the highest values)] and 14 samples located in the Al-Hawban sub-basin (wells No. 6, 12, 2, 11, 1, 36, 30, 25, 42, 29,24, 28, 55 and 34); however, there are 8 wells with water type of Na-Cl that have fluoride's concentration equalizer or

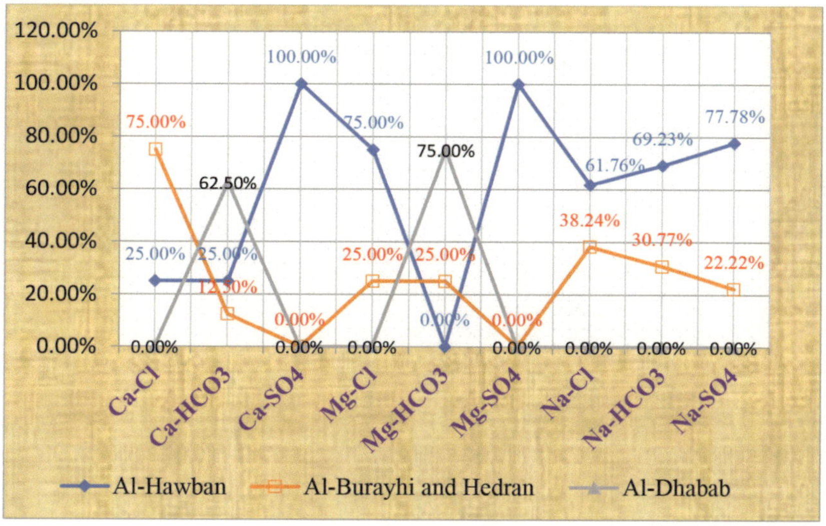

Figure 18.
The distribution of water type according to sub-basins of the study area.

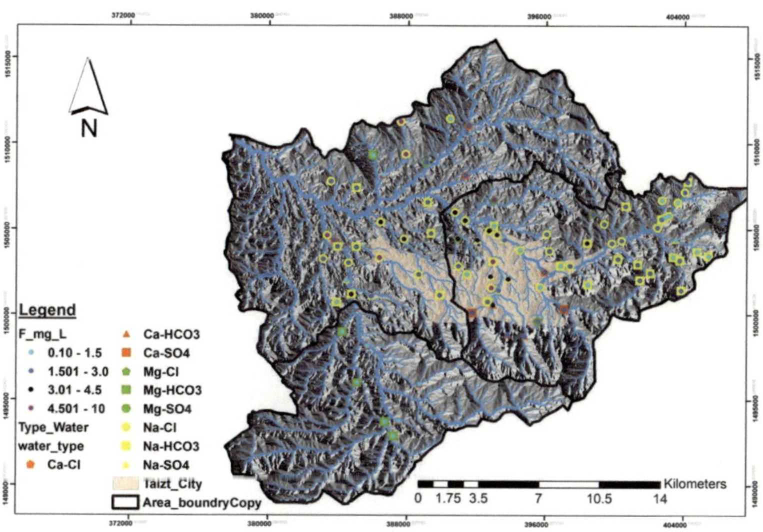

Figure 19.
The spatial distribution both of water type and the F- mg/L in the study area.

lower than 1.5 mg/L [(7 wells located in the Al-Hawban sub-basin (wells No. 54, 37, 49, 53, 51, 40 and 5) and one well (No. 72) located in the Al-Burayhi and Hedran sub-basin)]. The wells with the Na-HCO3 water type have 19 groundwater samples out of 26 samples (wells and one spring) with a high fluoride's concentration [(11 wells and one spring in Al-Hawban (dug wells No. 9, 22, 27, 18, 15, 47, 17, 19, 20, 23 and 52, spring No. 16) and 7 wells in Al-Burayhi and Hedran sub-basin (No. 67, 85, 70, 89, 69, 84 and 86)]. There are 7 samples out of 9 groundwater samples with water type of Na-SO4 have a high values of fluoride's concentration, 6 wells located in the Al-Hawban [(wells No. 39, 33, 4, 46, 3 and 35) and one located in Al-Burayhi and Hedran sub-basin (well No. 66)]. All samples with water type both of Ca-SO4, Mg-Cl, Ca-Cl and Mg-SO4 have abnormal fluoride's concentration as shown in **Table 5**. Abnormal concentration of fluoride was more prevalent with groundwater samples with Mg-Cl water type in Al-Hawban sub-basin (well No. 41 with

Water type	Normal concentration N (%)	Abnormal concentration N (%)	P-value
Mg-HCO3	5 (100%)	NIL	1
Ca-HCO3	5 (62.5%)	3 (37.5%)	0.3
Mg-SO4	Nil	1 (100%)	0.10
Na-HCO3	7 (26.9%)	19 (73.1%)	0.01
Na-SO4	2 (22.22%)	7 (77.78%)	0.02
Na-Cl	8 (23.53%)	26 (76.47%)	0.01
Ca-SO4	Nil	2 (100.0%)	0.05
Mg-Cl	Nil	4 (100%)	0.02
Ca-Cl	Nil	4 (100%)	0.02

Table 5.
Water type, normal and abnormal of F- mg/L and number of wells.

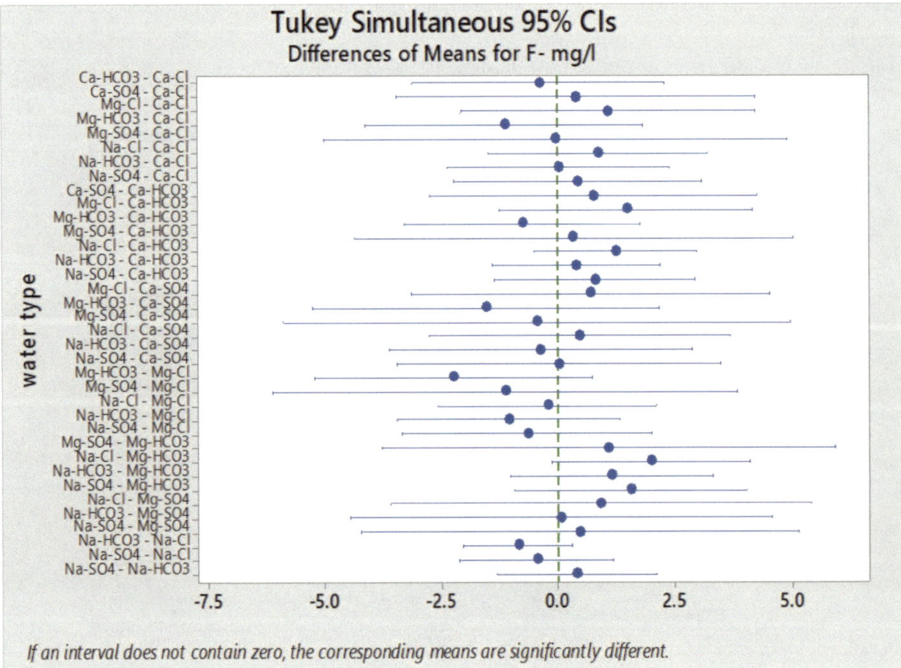

Figure 20.
The difference of means for F- mg/L with 95% Cl.

4.87 mg/L, No. 26 with 1.58 mg/L and well No. 10 with 3.33 mg/L) and Al-Burayhi and Hedran (source No. 81 with 2.66 mg/L); and with water type both of Ca-SO4 (wells No. 38 and 8 in Al-Hawban sub-basin), Ca-Cl (wells No. 71, 73 and 91 in the Al-Burayhi and Hedran sub-basin and well No. 31 in Al-Hawban sub-basin) and Mg-SO4 (well No. 7 in Al-Hawban sub-basin). **Figure 20** illustrated that there are no significant different in the fluoride concentration between the sources of groundwater samples according to the water type.

3.2.4 The spatial distribution both of water type and F- concentration according to pH

The distribution type of groundwater according to pH showed that, from Mg-So4 type through a Na-HCO3, Na-So4 and Na-Cl type groundwater to Mg-Cl and

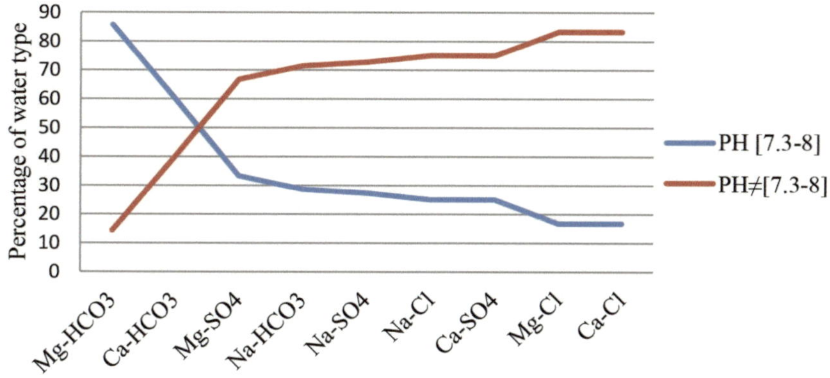

Figure 21.
Distribution of water type according to pH.

	ORcrude	CI for 95%	p-value	ORa	CI for 95%	p-value
pH [7.3–8]	0.417	0.164–1.062	0.067	0.373	0.139–0.99	0.050
Water type	1.755	1.295–2.378	0.000	1.711	1.261–2.322	0.001
Sub-basin	0.389	0.194–0.778	0.008	0.366	0.176–0.760	0.007

ORa: odds ratio adjusted. pH ≠ [7.3–8]: reference. Mg-HCO3 water type: reference. Al-Hawban sub-basin: reference.

Table 6.
Logistic regression univariate and multivariate model for fluoride variation.

ultimately Ca-Cl when the pH range between 7.3 and 8 was more lower than pH ≠ [7.3–8] (**Figure 21**) and the comparison of the groups "pH [7.3–8]" and "pH ≠ [7.3–8]" showed a significant difference in fluoride concentration ($p < 0.05$).

Results of the multivariate analysis are shown in **Table 6**. In the two logistic regression model, after adjusting for pH, water type and sub-basin (ORa = 0.366; CI: 1.76–0.76), the water type (ORa = 1.71; CI: 1.261–2.32) remained dependently associated with abnormal fluoride concentration (**Table 6**).

3.2.5 The relationship between the concentration of fluoride and type of sources of samples

The results statistic analysis show that the different of the fluoride concentration between the different sources of water samples (dug well, bore well and spring) in the study area is not significantly different. Fluoride concentration decreases not significantly according to well's type $F(2) = 2.19$, $p = 0.121$.

Fluoride concentration is positively and not significantly related to depth of the groundwater ($r = 0.046$, $p > 0.05$).

3.2.6 The comparison of the abnormal and normal fluoride concentration according to TDS

The comparison of the abnormal and normal fluoride concentration according to TDS (total dissolved solids) in the three sub-basins showed a significant differences between the three sub-basins ($p < 0.0001$) and positive relationship between the fluoride concentration and TDS ($r = 0.5$; $p < 0.0001$). A multiple comparison of median concentration among these sub-basins in fluoride shows that Al-Hawban and Al-Burayhi and Hedran sub-basins reach higher fluoride content which is more

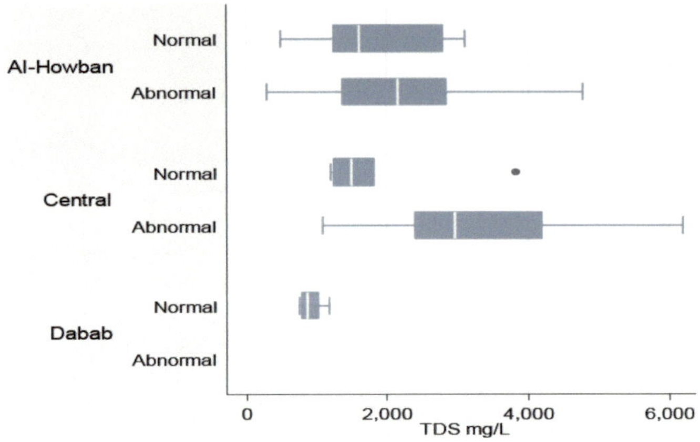

Figure 22.
Box plot for the maxi, min and average of the fluoride content in groundwater according to TDS of three sub-basins.

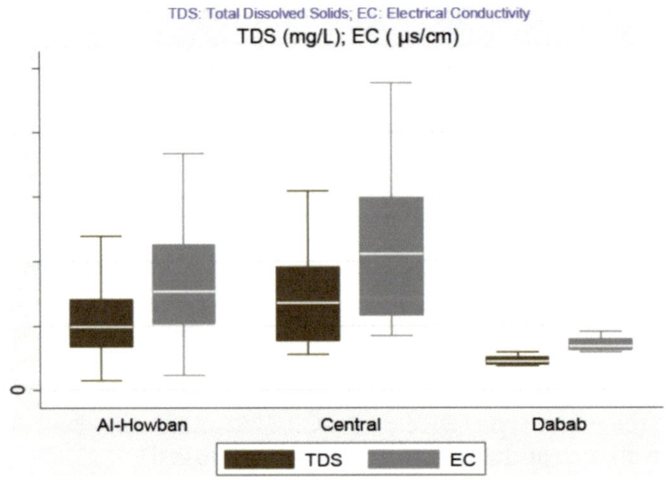

Figure 23.
Box plot for the max, min and average of electrical conductivity and total dissolved solids according to study sub-basins.

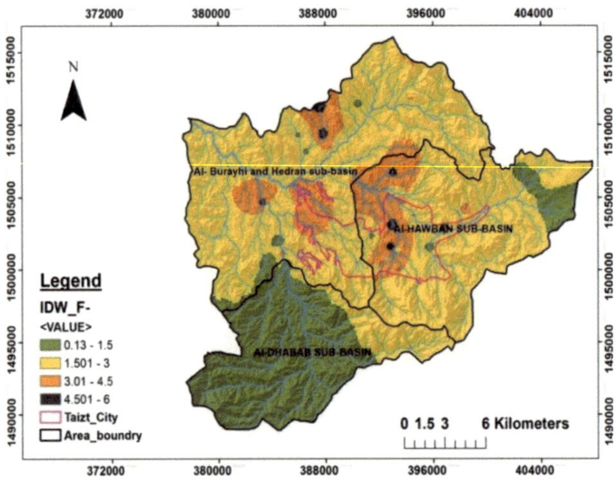

Figure 24.
Spatial distribution of fluoride concentration in the study area.

than 1.5 mg/L and similar but significantly higher than the Al-Dhabab sub-basin; Al-Hawban-Al-Burayhi and Hedran: ($p > 0.072$); Al-Hawban-Al-Dhabab: ($p < 0.003$); Al-Dhabab-Al-Burayhi and Hedran: ($p < 0.0001$). Moreover, this increase in TDS has been found only at two sub-basins when the fluoride is abnormal content (**Figure 22**). The results of the study samples of electrical conductivity and TDS according to study sub-basins are given in **Figure 23**.

3.2.7 Spatial distribution of fluoride

In order to enable sustainable development of groundwater resources, it is necessary to delineate the safe and unsafe zones with reference of fluoride content (1.5 mg/L); hence spatial distribution of fluoride's concentration was mapped in the three sub-basins of the study area (**Figure 24**).

Based on the Kruskal-Wallis test for the various sub-basins, the level of significance was ($p < 0.05$). This result illustrated that there were a significant differences between three sun-basins with regard to the concentration of fluoride in groundwater. The differences of means for fluoride concentration in groundwater for the three sub-basins show the mean's fluoride in Al-Burayhi and Hedran (F-) > Al-Hawban (F-) > Al-Dhabab (F-). A multiple comparison of mean concentration of fluoride among these sub-basins shows that there is no a significant difference between the Al-Hawban and Al-Burayhi and Hedran sub-basins P-value 0.277, there is a significant difference between the Al-Hawban and Al-Dhabab sub-basins P-value 0.017 and there is a significant difference between the Al-Burayhi and Hedran and Al-Dhabab sub-basins P-value 0.001.

4. Discussion

Fluoride concentration variation is widely in the study area from 0.1 mg/L (in well No.93 in Al-Burayhi and Hedran sub-basin) to 6 mg/L [in Well No. 83, of the same sub-basin (Al-Burayhi and Hedran sub-basin)]. We observed that the concentration of fluoride in the Al-Dhabab sub-basin is the optimal concentration according to the WHO drinking water guidelines value of 1.5 mg/L.

Waters with high fluoride concentrations occur in large and extensive geographical belts associated with (a) sediments of marine origin in mountainous areas, (b) volcanic rocks and (c) granitic and gneissic rocks [21], and the high concentration of fluoride widely accepted that most of the F are derived mainly from acidic volcanic rocks such as pumice, obsidians, pyroclastic deposits, ignimbrites and rhyolite, and the main minerals for F are fluorite, fluorapatite, micas and hornblende [22]. Because the geology of study mainly constituent from the acid and basic volcanic and grants rocks, the level of fluoride concentration in the Al-Dhabab sub-basin can be explained by the nature on aquiver in this study area (Cretaceous Tawilah Sandstone), while the groundwater in the other sub-basin is produced either from Tertiary fractured volcanic (that have F- bearing mineral and the groundwater in this aquiver have long-time contact with aquiver, which adjudge the important factors leading to the high fluoride concentration result of interaction between the groundwater and the aquiver) or from the Quaternary alluvium aquiver, where the Wadi sediments deposited are derived from the alteration of volcanic ash and tufa mainly of (Tr1); this quiver depends on their recharge mainly on the wastewater of the urban and industrial activates; this aquiver exposed to over exploration of their groundwater and finally the dry and semi-dry condition plays an important role in the degradation of groundwater in this aquiver.

It is clearly observed that the Al-Burayhi and Hedran and Al-Hawban sub-basins have the highest concentration of fluoride ion in the chemistry of water. Highest concentrations were found to be 6 mg/L from Al-Burayhi and Hedran sub-basin, 5.81 mg/L from Al-Hawban unlike the Al-Dhabab sub-basin which remains unaffected by the contamination fluoride of groundwater. According to the report of [23], the dental fluorosis is the widely fluoride disease observed in the affected areas, and there is a positive relationship between fluoride in water and the occurrence of dental fluorosis in Taiz region.

In order to understand the vertical distribution of the fluoride ion concentration from the water of the study area, the type of the sample water (dug well, bore well and spring) evaluated separately. There was no significant difference between the three well types, dug well sample, springs and bore wells. It can be concluded that shallow aquifers do not reflect higher fluoride contamination than deeper aquifers. It is observed that most of the water samples showed enhanced concentrations with generally increasing trends to the low elevated area (Al-Burayhi and Hedran sub-basin), while the high elevation shows low concentration of fluoride (Al-Dhabab sub-basin). All the water samples collected from the uphill zones of Al-Dhabab sub-basin were exhibited low fluoride concentration.

Compared with Na-HCO3 type groundwater, Ca-HCO3 type groundwater is known to generally contain lower fluoride [24]. Its hydrochemistry is characterized by increased Ca2+ ion concentration with increasing total dissolved solid due to the gradual dissolution of carbonate minerals or Ca2+ bearing plagioclase in aquifer materials [25, 26]. The Na-HCO3 type groundwater is generally enriched in fluoride and sodium ions, due to the dissolution of silicates as well as the removal of Ca2+ by calcite precipitation and cation exchange [27, 28]. The solubility limits for fluorite and calcite provide a natural control on water composition in a view that calcium, fluoride and carbonate activities are interdependent [29, 30]. In addition to the effect of those areas by different liquid waste by runoff and sewage disposal, the heavy pumping of well water is also contributed because of the scarcity of water which leads to the increase of the concentration of salt in the water. TDS levels ranged widely from 291 to 6188 mg/L with most station levels above 400 mg/L and many of the samples studied were higher than the permissible limit of 1500 mg/L according to WHO (2003). This wide variation in TDS values indicates that the area hydrochemistry is influenced by diverse processes such as water-rock interaction and anthropogenic pollution. Fluoride concentrations frequently are proportional to the degree of water-rock interaction because fluoride primarily originates from the geology [9, 31–34]. Due to the high rainfall, rugged topography, factories, lack of total coverage per sewerage network, population density and faults in the study areas could also explain this high fluoride content by runoff and infiltration of chemical fertilizers in agricultural areas, septic and sewage treatment system discharges with fluoridated water supplies and liquid waste from industrial sources. The topography of the study areas varies from level plain to steep slopes. Study area ranges in elevation between 900 and 3000 m above sea level. Taiz plain receives about 500 mm/year of rainfall and significant recharge form runoff of surrounding mountains [35]. In addition to this groundwater fluoride pollution that can affect human health, there have been indications that uptake of fluoride from other sources like food, dust and beverages may be many times higher than that of water [36]. About the fluorosis in selected villages of Taiz Governorate, the percentage of children with fluorosis was very high. Not only because of drinking water, various food habits (like drinking black tea and Chewing Qat) indicated a high contribution of fluoride to food. In AL-Hawban sub-basin, some of children, especially from Jabal Sabir area, used to chew Qat daily, and the Qat are cultivated in the man-made terraces of Jabal Sabir alkali granite, where it expected to be the main source of F- reach minerals like fluorite [23].

On the other hand, the use of fluoridated water for cooking increases the fluoride content significantly especially in dry foods like maize flour which absorbs much water during cooking. It has been reported that fluoride availability may be influenced by simultaneous intake of food and fluoride containing compounds in a positive or negative manner depending on the food type, mode of administration and type of fluoride compound [37].

5. Conclusion

Much of the fluoride entering the body is from water and the high concentration of fluoride in water's sources is therefore a major concern. The fluoride is found in the atmosphere, soil and water. It enters the soil through weathering of rocks, precipitation or waste runoff. Understanding of the fluoride occurrence is important in the management of the fluoride related epidemiological problems. Al-Hawban and Al-Burayhi and Hedran are the worst sub-basins affected by fluoride contamination in drinking water. 71% of samples (66 samples out of 93 samples) in the study area have F- concentration (mg/L) above the permissible limit and alternate water sources will be difficult. Therefore, defluoridation of drinking water is the only practicable option to overcome the problem of excessive fluoride in drinking water in these areas. More refined studies however need to be done before any long-term intervention efforts can be planned. In the meantime, there is a critical need to educate young Yemenis about fluorosis and simple intervention measures to avoid long-term health problems. Other studies in the region are urged studying the cause and effect relationship between the abnormal content fluoride and population health.

Author details

Ramzy Saeed Mahbob Naser[1,2]*, Mohammed El Bakkali[3] and Driss Belghyti[4]

1 Al-Hodeidah University, Al-Hodeidah, Yemen

2 Ministry of Public Works and Roads, Sana'a, Yemen

3 National Mutual Aid, Kenitra, Morocco

4 Ibn Tofail University, Kenitra, Morocco

*Address all correspondence to: ramzi_mh1@yahoo.com

References

[1] Guo Q, Wang Y, Teng Ma RM. Geochemical processes controlling the elevated fluoride concentrations in groundwaters of the Taiyuan Basin, northern China. Journal of Geochemical Exploration. 2007;**93**:1-12

[2] Tahaikt M, Ait Haddou A, El Habbani R, Amor Z, Elhannouni F, Taky M, et al. Comparison of the performances of three commercial membranes in fluoride removal by nanofiltration, continuous operations. Desalination. 2008;**225**: 209-219

[3] Bell MC, Ludwig TG. The Supply of Fluoride to Man: Ingestion from Water. Fluoride and Human Health. WHO Monographs. Geneva: World Health Organization; 1970

[4] Sorg TJ. Treatment technology to meet the interim primary drinking water regulations for inorganics. Journal of American Water Works Association. 1978;**70**(2):105-111

[5] Mahramanlioglu M, Kizilcikli I, Bicer I. Adsorption of fluoride from aqueous solution by acid treated spent bleaching earth. Journal of Fluorine Chemistry. 2002;**115**:41-47

[6] Susheela A, Kumar AK, Bhatnagar M, Bahadur M. Prevalence of endemic fluorosis with gastro-intestinal manifestations in people living in some north-Indian villages. Fluoride. 1993;**26**: 97-104

[7] Nanyaro JT, Aswathanarayana U, Mungere JS, Lahermo P. A geochemical model for the abnormal fluoride concentrations in waters in parts of northern Tanzania. Journal of Asian Earth Sciences. 1984;**2**:129-140

[8] Gaciri SJ, Davies TC. The occurrence and geochemistry of fluoride in some natural waters of Kenya. Journal of Hydrology. 1993;**143**:395-412

[9] Gizaw B. The origin of high bicarbonate and fluoride concentrations in waters of the main Ethiopian Rift Valley. Journal of African Earth Sciences. 1996;**22**:391-402

[10] Subba Rao N, John DD. Fluoride incidence in groundwater in an area of peninsular India. Environmental Geology. 2003;**45**:243-251

[11] Gupta SK, Deshpande RD, Agarwal M, Raval BR. Origin of high fluoride in groundwater in the North Gujarat-Cambay region, India. Hydrogeology Journal. 2005;**13**:596-605

[12] Jacks G, Bhattacharya P, Chaudhary V, Singh K. Controls on the genesis of high-fluoride groundwaters in India. Applied Geochemistry. 2005; **20**:221-228

[13] WHO. Guidelines for Drinking-Water Quality. 3rd Ed. Geneva: World Health Organization; 2006. ISBN: 92-4-1546743

[14] KS V. Fluorosis in Yemen-Prevention-Control Status and Strategies for Mitigation. Sana'a, Yemen: Natural Water Resources Authority (NWRA); 2008 (unpublished report)

[15] Naser R, El Bakkali M, Darwesh N, El Kharrim K, Belghyti D. Assessment of groundwater quality in the Al- Burayhi and Hedran sub-basin, Taiz, Yemen - a GIS approach. Dans: E3S Web of Conferences. 2018;**37**:01005

[16] UNICEF. Survey Report About the Effect of Fluoridation Among School Children in the District of Sanhan; 2008.

[17] Naser R, El Bakkali K, AL Nahmi F, Darwesh N, El Kharrim K, Belghyti D. Environmental contaminants and their impact on groundwater quality, in three area of Taiz, Yemen. Rev L'ENTREPRENEURIAT

L'INNOVATION [En ligne]. 2017;**1**(4): 1–6. Available from: https://scholar.google.com/scholar?hl=ar&as_sdt=0%2C5&q=Environmental+Contaminants+and+Their+Impact+on+Groundwater+Quality%2C+in+three+area+of+Taiz%2C+Yemen&btnG=

[18] Kader AAAA. Geophysical and Hydrogeological Studies for Exploration and Evaluation of the Groundwater and its Environmental Protection in the Uplands of Wadi Rasyan, Taiz, Yemen. Taiz: Assiut University; 2005. p. 203

[19] El-Yemen D. Hydrogeological and Land-Use Studies of the Ta'iz Region (Upper Wadi Rasyan Catchment). Sana'a: UNDDSMS; 1997

[20] Alenbeaaoy Alkherbash. Geology of Yemen. Dar Aobadi for Publishing and Distribution; 1996

[21] Chilton J, Dahi E, Lennon M, Jackson P. Fluoride in Drinking-water. IWA Publishing: WHO; 2006

[22] Al-mikhlafi AS. Groundwater quality of Yemen volcanic terrain and their geological and geochemical controls. Arabian Journal of Geosciences. 2010;**3**:193-205. DOI: 10.1007/s12517-009-0068-7

[23] NWRA. Study about the Fluorosis in Selected Villages of Ta'aiz Governorate. Sana'a, Yemen: Natural Water Resources Authority (NWRA); 2009 (unpublished report)

[24] Lee JU, Chon HT, John Y. Geochemical characteristics of deep granitic groundwater in Korea. Journal of Soil and Groundwater Environment. 1997;**4**:199-211

[25] Yun ST, Koh YK, Choi HS, Youm SJ, So CS. Geochemistry of geothermal waters in Korea: Environmental isotope and hydrochemical characteristics. II. Jungwon and Munkyeong areas. Economic and Environmental Geology. 1998;**31**:201-213

[26] Yun ST, Chae GT, Koh YK, Kim SR, Choi BY, Lee BH, et al. Hydrogeochemical and environmental isotope study of groundwaters in the Pungki area. Journal of Soil and Groundwater Environment. 1998;**5**: 177-191

[27] Chae GT, Yun ST, Choi BY, Kim K, Shevalier M. Geochemical concept and technical development of geological CO2 sequestration for reduction of CO2. Economic and Environmental Geology. 2005;**38**:1-22

[28] Kim K, Jeong G. Factors influencing natural occurrence of fluoride rich groundwaters. A case study in the southeastern part of the Korean peninsula. Chemosphere. 2005;**58**: 1399-1408

[29] Kim JY, Chon HT. Pollution of a water course impacted by acid mine drainage in the Imgok creek of the Gangreung coal field, Korea. Applied Geochemistry. 2001;**16**:1387-1396

[30] Kundu N, Panigrahi MK, Tripathy S, Munshi S, Powell MA, Hart BR. Geochemical appraisal of fluoride contamination of groundwater in the Nayagarh district of Orrissa India. Environmental Geology. 2001;**41**: 451-460

[31] Banks D, Reimann C, Røyset O, Skarphagen H, Sæther O. Natural concentrations of major and trace elements in some Norwegian bedrock groundwaters. Applied Geochemistry. 1995;**10**:1-16

[32] Dowgiałło J. Thermal water prospecting results at Jelenia Góra-Cieplice (Sudetes, Poland) versus geothermometric forecasts. Environmental Geology. 2000;**39**: 433-436

[33] Frengstad B, Banks D, Siewers U. The chemistry of Norwegian groundwaters: IV. The dependence of element concentrations in crystalline

bedrock groundwaters. Science of the Total Environment. 2001;**277**:101-117

[34] Carrillo-Rivera JJ, Cardona A, Edmunds WM. Use of abstraction regime and knowledge of hydrogeological conditions to control high-fluoride concentration in abstracted groundwater: San Luis Potosí basin, Mexico. Journal of Hydrology. 2002;**261**:24-47

[35] Abdulaziz A. Geophysical and Hydrogeological Investigations and Groundwater Quality Evaluation and Protection from Upper Wadi Rasyan. Taiz: Yemen. Assiut University; 2005

[36] Chowdhury NG, Brown RH, Shephered MG. Fluoride intake by infants in New Zealand. Journal of Dental Research. 1990;**69**(12):1828-1833

[37] Trautner K, Einwag J. Influence of milk and food on fluoride bioavailability from NaF and Na2FPO3 in man. Journal of Dental Research. 1989;**68**(1):72-77

Chapter 10

Oil–Water Separation Techniques for Bilge Water Treatment

Nurul Aini Amran and Siti Nor Adibah Mustapha

Abstract

Discharging accumulated bilge water from the ship is very important in order to maintain its stability and safety. However, the bilge water that contains contaminants, including waste oils and oily wastes, must be treated prior discharging to the sea. The International Convention for the Prevention of Pollution from Ships (MARPOL) has set strict oil discharge limit in order to minimize sea pollution. Thus, an efficient oil–water separator must be installed to separate the oil from the bilge water. This chapter introduces and discusses the working mechanisms, as well as advantages and disadvantages of the available oil–water separation techniques for bilge water treatment, which include gravitational, centrifugation, flotation, coagulation and flocculation, biological processes as well as absorption and adsorption.

Keywords: bilge water, oil/grease, oil–water separation, centrifugation, flotation, coagulation and flocculation, biological

1. Introduction

Shipping has a vital role in developing human society over the years, at which the shipping activities have linked the widely separated parts of the world through commercial relationships. In fact, the shipping industry is still developing from time to time with rapid industrial and digital economy growth. In Canada, the shipping industry has been established since 1840 and now undergone significant technological advances, where the size of ships that carry containers for international use continues to increase [1]. Even in Malaysia, the government has launched an initiative, namely, Malaysia Shipping Master Plan, that ensures the shipping industry is focused on developing itself and has a guide for future development. This initiative takes place from 2017 until 2022 with a "Revitalizing Shipping for a Stronger Economy" theme [2].

Generally, there are three main classifications for the global cargo shipping industry, which are:

i. Wet bulk: Transportation of crude oil and other petroleum products

ii. Dry bulk: Shipment of bulk goods

iii. Liners: Small shipments of general commercial goods

Each of wet bulk, dry bulk, and liner shipping needs their specialized vessels, which are tankers, bulkers, and container ships, respectively [3].

Firstly, for wet bulk shipping, approximately a quarter of the goods transported by sea is dominated by crude oil [4]. The oils are transported from its production point to the purchasers by the wet bulk shipping or known as tankers. Majority of the crude oil is moved from the most significant oil-producing region, which is the Middle East, to the dominant importers like the European Union, Japan and the United States of America. Other than that, North America imports oil from the Caribbean and West America meanwhile West and North Africa export their oils to Europe [4].

Next, the largest group out of these three classifications is dry bulk shipping, where more than 50% of all loaded goods are handled by the bulkers, while 30 and 16% are for tankers and containers, respectively [5]. There is a vast range of solid cargoes transported by containers. Generally, there are five primary crucial bulk goods, which are coal, grain, iron ore, bauxite, and phosphates, in which iron ore and coal are the two goods that are transported the most [4]. Meanwhile, chemical packages and steel products are the example of the shipped minor bulk goods [6]. The main routes for iron ore transportation are from Australia and Brazil to Japan as well as from Brazil to Western Europe. While for coal, which is commonly used as steam coal in power stations to generate electricity, the leading exporters are from South Africa, Australia, Colombia, the East and West Coast of the United States, as well as Indonesia. Moreover, Australia, South Africa, Colombia, and the East Coast of the United States export their coal to Western Europe, whereas Japan receives the coal from the West Coast of the United States, Australia, and South Africa [4].

Liner shipping, also known as container shipping, provides services by transporting goods in containers with scheduled sailings. The variety of goods transported by liner shipping are packed in a smaller unit. One of the contributors to the continuation and development of liner shipping is the increment of the digital economy. According to Ref. [7], companies that produce and process raw materials, commodities, and manufacturing goods are the previous world's fastest-growing and biggest corporations. However, currently, Internet-related service and technology-based manufacturers, such as Alibaba, Amazon, Apple, and Microsoft, have become the world's most valuable and most prominent companies, where e-commerce, online communication and cashless Business to Business (B2B) and Business to Consumer (B2C) transactions are practiced [7]. This development has contributed more to the growth of the shipping industry, where it is considered as a catalyst for economic development by facilitating world trade, due to the cheaper mode of transportation.

However, aside from on-the-ground activities such as lubricants, refineries, and petrochemical industries, it is undeniable that shipping activities have contributed to marine pollution, especially in this twentieth century where carriage of the cargoes by the ships is increasing. According to [8], millions of tons of oily wastes and waste oils are generated as the by-products of the ships, every year. One of the contributors to water pollution by the operating marine vessels is the discharging of oily bilge water. Typically for marine vessels, the oily wastes and waste oil that come from various sources accumulate in the bilge space, which is the lowest part of the vessel.

Routinely, the accumulated oily bilge water must be discharged out of the bilge spaces to maintain the stability of the vessel, hence eliminating the possibility of the ship to be in the conditions that can cause a hazard to it [9]. The wastes discharged can eventually cause water pollution, which leads to many negative impacts on the human, environment, and marine populations. Oily bilge water may poison marine organisms because it might cover plants and tiny animals when it floated on the surface of the water and is carried into the shoreline, causing life cycles of the plant and the respiration of the animals be interfered [10].

Hence, many governments and international industries are working on the marine pollution issue, mainly originated from the shipping industry, such as the Marine Environmental Protection Committee (MEPC), the International Convention for the Prevention of Marine Pollution from Ships (MARPOL), and the Department of Environment (DOE), Malaysia. In order to solve the problem, oily wastewater separator is essential and needs to be installed and operated effectively to prevent the pollution as well as to ensure that the water discharged overboard is within legal limits.

The lowest compartment of the vessels and directly above the keel is known as bilge, where water that drains off from various sources is captured. The water might be originated from rain, interior spillage, rough seas, or minor leakage in other main parts of the vessel. Depending on a few factors such as ship size, design of engine room, and components' age, the amount of accumulated bilge water onboard varies from one to another. In order to maintain the stability of the vessel and to avoid conditions that can cause hazard (such as vessel's propulsion systems and ancillary machinery damage and fire hazard) due to too much of bilge waste accumulation, it is crucial to remove the bilge water into a holding tank, periodically [9]. There are two options to manage the bilge water, which are whether installing bilge separator to treat it onboard or holding it in a tank on the vessel before discharging it to the shore's treatment facility. Somehow, treating it onboard has an advantage where a smaller volume of oily bilge water has to be stored in the vessel. Meanwhile, the treated wastewater can be removed according to the related regulations and standards.

The composition of bilge water depends on the design and function of the ship. The wastewater is commonly comprised of water, oily fluids, cleaning fluids, lubricant, and grease as well as other wastes that originated from piping, engines, and other operational and mechanical sources in the vessel's machinery spaces [9] as well as urine and chemicals. Other than that, extra waste streams in massive vessels contain sludge, waste oil, and oily water mixtures. Sludge is formed from the continuous fuel purification to remove contaminants in order to enhance low-quality fuels as well as to avoid ship's engines and highly machined components from being damaged.

2. Impact of bilge on marine pollution

The regulation stipulated by the International Maritime Organization (IMO) highlighted on the oil content of the bilge water discharged to the sea. The International Convention for the Prevention of Pollution from Ships (MARPOL 73/78) has set the maximum limit of 15 mg/L for the oil content in the wastewater to be discharged to the sea. According to the US EPA (2008), passenger ships produce the highest amount of bilge water with huge difference as compared to the other types of ships. This is due to their more complex constructions and support for crowds of passengers [9].

Typically, the small volumes of treated bilge water are released above the water line and instantly diluted in the sea water. Hence, the obvious effects of oil spill is most likely not going to occur. However, a long-term effect might happen to the marine living organisms around the shipping lanes. The negative consequences that will take place may be due to the excessive concentration of biodegradable compounds, including oil, as well as continuous increment of nondegradable compound concentration such as metals [11].

Other than that, surfactant is one of the significant chemicals contained in the bilge water. The mixtures of oil and surfactants may cause higher toxicity since the oil and surfactants alone are toxic themselves. This may be due to the synergistic effects or the crude oil that has been dissolved, causing it to be consumable for the exposed organisms [12–15].

3. Current oil–water separation techniques

Typically, OWS is made up of three segments, which are separator unit, filter unit, and oil content monitor and control unit. The separator and filter units are included as treatment units, where many designs and different principles are applied. The gravity and centrifugal separators are commonly used as the first stage of the treatment, followed by other separation techniques, which is called as polishing treatment. The examples of the polishing unit are flotation, coagulation and flocculation, filtration, biological treatment, as well as absorption and adsorption [16]. Normal techniques, such as gravitational and centrifugation, are used for oily wastewater that has two distinct phases; meanwhile, addition of chemical or biological de-emulsification is required for separation of emulsified oily bilge water [17].

3.1 Gravitational method

Typically, oily bilge water treatment onboard starts with a gravitational method in order to remove heavy fractions and lighter fractions based on density difference. In this method, coalescing materials made of oleophilic polymer in the form of loose-packed media or parallel plate are used to attract the oil droplets to adhere to the plate [16]. Examples of oleophilic polymer used as the coalescing plate separators are polyethylene, fiberglass, and nylon [18].

The free-moving dispersed oil droplets continue to adhere to the plate or media until it can break from the coalescing material and float up to the surface of the tank. The presence of the oil detected by the sensors then automatically triggers the OWS to remove the collected oil to a waste oil tank. However, this method can only be effective when the phase of the oil and water is separated distinctively [19]. In other words, in some instances, the gravitational method is not suitable since the bilge water typically consists of emulsified oil formed due to the chemical emulsifiers (solvents and cleaning agents) as well as mechanical means such as ship's motion and transfer system pump [16]. **Figure 1** shows the gravitational separator process.

As can be seen from the figure above, as the oily bilge water flows through the parallel plate, oil globules are formed and float up to the surface to form oil layer. Oil skimmer is used to skim off the oil layer. Then, oil discharge valve and purge water valve are opened, where the oil is removed from the unit by the purge water.

3.2 Centrifugation method

Centrifugal separators are the alternative option for the gravity separators. The same principle is applied, in which the oil is separated based on the different density of oil and water as well as coalescence of the oil droplets. Somehow, the centrifugal acceleration causes the gravity to increase more, and the coagulation and flocculation processes are enhanced in order to separate the emulsified oil. This type of separator has many advantages as compared to gravity OWS. Since it can separate more oil from the bilge water, including emulsified oil, less oil content is loaded to the next treatment step, which is usually called a polishing unit. Thus, the polisher's

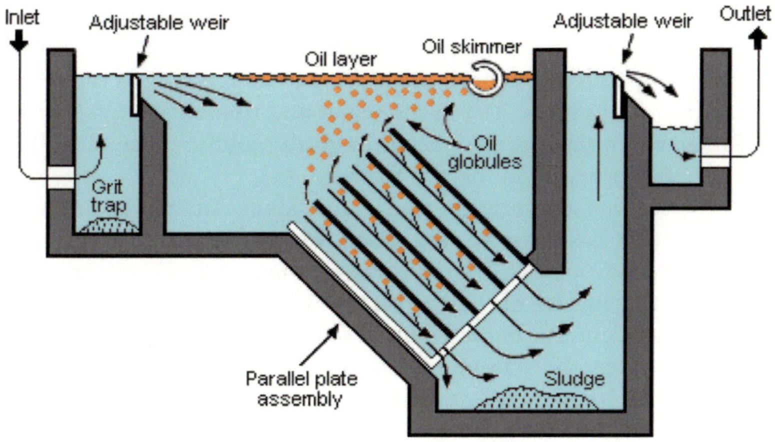

Figure 1.
Gravitational separator [20].

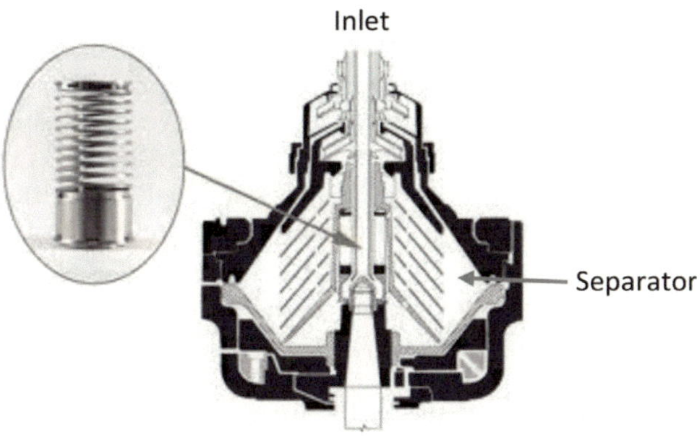

Figure 2.
Patented Alfa Laval XLrator [21].

service life might last long, reducing the cost of maintenance and repair. Centrifugal separators are also more compact and require smaller bilge water holding tanks [16]. However, high capital cost is needed for the centrifuges, and regular maintenance must be done since large horsepower motors are used during the process.

Figure 2 shows an example of manufactured separator (PureBilge by Alfa Laval).

In the inlet stream of the unit, the bilge water is accelerated by the XLrator with less shearing and foaming in order to prevent the oil drops from separating and further emulsion formation. Then, it flows into the separator, in which coalescence occurs due to high centrifugal force. Flocculation of small oil drops takes place and flocculants is added to promote bigger flocs for easier separation [21].

3.3 Flotation

Separation of oil by flotation occurs due to the difference in density of oil and water, where water is denser than oil, forming a scum layer on top of the water. Floatation technique can be divided into many different techniques, including electroflotation, froth flotation [22], and dissolved air flotation (DAF). Electroflotation separates oil from water through electrochemical reactions by electrolysis, where tiny bubbles produced from electrolysis will cause the pollutants to float to the water body surface [23]. In froth flotation, the separation takes place when the oil adheres onto the fine bubbles generated when air is introduced into the system. Surfactant is added to adsorb the air or water interface of the bubbles of air with the head groups (hydrophilic) in the water and the tail groups (hydrophobic) in the air. Hence, when the bubbles rise through the solution, the oil will concentrate on the bubble surfaces and foam is formed [24].

Meanwhile, dissolved air flotation (DAF) introduces micro gas bubbles into the flotation chamber that has been formed when water is saturated with gas under pressure [25]. The oil droplet will spread around the gas, and conglomerate will continue to rise to the surface of the solution. The advantages of flotation treatment are the following: less investment needed, low energy consumption, and easy to maintain [26]. However, the statement contradicts with Yu et al. [27] who stated that flotation requires high energy consumption and has repairing and maintenance problem as well as issue in manufacturing of the device.

3.4 Coagulation and flocculation

According to Yu et al. [27], coagulation process is a robust oil–water separation technology because it is able to separate dissolved and emulsified oil; hence, it is vastly applied in the latest oily wastewater treatment method [28]. In the coagulation process, coagulant, a chemical substance, is added to the wastewater to destabilize the charge of colloidal particles in the solution [29] which is too tiny for gravitational settling. When the particles are destabilized, larger flocs are formed making it easier to settle and then are skimmed off to the clarifier or sludge thickener.

Anyhow, many experiments might be needed if the coagulation method is going to be used in treating oily wastewater. This is because of the complexity of oily wastewater, making it hard to choose the most suitable coagulants for effective separation of oil and water to take place [27]. In a study done by Zeng et al. [30], oil removal efficiency is improved up to 99% when aggregation of poly-zinc silicate (PZSS) with anionic polyacrylamide (A-PAM) is used as the coagulating and flocculating chemicals. Somehow, higher costs are needed, and it could cause water bodies' secondary pollution and difficulties to the next process [27].

3.5 Biological

Some bilge water treatment units include biological treatment, which is called as a bioreactor. In this method, microorganisms are used to eliminate or reduce the organic and inorganic compounds before the treated wastewater being discharged to the sea or to a collection system [29]. The microorganisms convert the dissolved and particulate carbonaceous organic matter, including oil, in the bilge water into simple end products through the oxidation process. The equation below is representing the aerobic biological oxidation of organic matter [29].

$$\text{Organic material} + O_2 + NH_3 + PO_4^{3-} \rightarrow \text{new cells} + CO_2 + H_2O \tag{1}$$

As can be seen from Eq. 1 above, the oxidation process needs oxygen (O_2) and nutrients, which are nitrate (NH_3) and phosphate (PO_4^{3-}), in order to convert the organic matter to carbon dioxide (CO_2) and water (H_2O). Other than that, the new cells generated are referring to the biomass produced after the oxidation of organic matter takes place.

There are two principals of biological processes used to treat wastewater, which are suspended growth and attached growth, or also known as biofilm, processes.

Suspended growth process maintains the microorganisms in liquid suspension by proper mixing methods. Meanwhile, attached growth attach the microorganisms to an inert packing material, where the wastewater will flow past the biofilm to remove the organic materials [29]. For OWS onboard, biofilm is used, in which the bacteria are attached to a synthetic support media. In this bioreactor, aerators are installed under the media to supply oxygen to the bacteria for bacterial growth as well as for oxidation of the organic contaminants to take place. Other than that, a clarifier is needed in order to remove the biomass formed at the end of the processes [16].

Biological treatment, with flexible operation, simple maintenance and management, as well as stable effluent quality [26], is indeed a suitable method since no waste oil is produced by the process [16]. Small oil droplets of emulsified oil, which are hardly removed by physical and chemical treatment, can be degraded easily by the bacteria [31]. However, CO_2 will be produced from the process, resulting in increment in greenhouse gases. Even though the operating cost is low, the capital

Techniques	Advantages	Disadvantages	References
Gravitational	Effective for discrete phases of oil and water	Not effective to separate emulsified oils from water	[16, 19]
Centrifugation	1. More compact 2. Do not require large bilge water holding tanks 3. Produce small amount of waste 4. Can be run without continuous man-hours operation and supervision	1. Use huge horsepower motors 2. Require frequent maintenance 3. High capital cost for centrifuges	[16]
Flotation	1. Less investment needed 2. Low energy consumption 3. Easy to maintain 4. Produce less sludge	1. Repairing and maintenance problem 2. Issue in manufacturing of the device	[26, 27]
Coagulation and flocculation	1. Can remove emulsified oil and dissolved oil 2. High adaptability	1. Need to be operated by skillful operator 2. High operating cost 3. Produce high amount of sludge which then needs to be disposed 4. A lot of experiments might be needed due to complexity of wastewater composition	[16, 27, 28, 33]
Biological	1. Able to effectively degrade organic pollutants including emulsified oil 2. Can remove other organic pollutants 3. No waste oil produced 4. Mechanically simple 5. Low operating cost	1. Loading spikes can occur 2. High capital cost 3. Need to be operated by skillful operator	[16, 27]
Absorption and adsorption	1. Suitable for less than 400 GT vessels 2. Relatively compact 3. Low capital and operation costs 4. Require low maintenance	1. Need to replace the media frequently (if necessary)	[16]

Table 1.
Comparison of the separation techniques.

cost needed is high and skillful personnel to be in charge is required to ensure the treatment functioned well.

Basically, there are three stages involved in the process [32]. The first stage is heavy phase separation, where separated oils and solids are removed. In the second stage, the bacteria are used to convert the emulsified oil into non-harmful end products. This stage is known as emulsified oil degradation stage. Then, the third stage contains clarifier in order to remove the remaining solids and water.

3.6 Absorption and adsorption

Physicochemical sorption involved in absorption and adsorption processes can eventually separate the oil from the bilge water. Absorption occurs when two discrete physical states of substances are fused together. Meanwhile, in adsorption, molecules adhere onto the surface of different phase [16]. Both absorption and adsorption involved absorbent sorption media and adsorbent, respectively. The oil is separated by pumping the bilge water through these media until their maximum sorption capacity is achieved, which then the oil is removed. Certain used media are possible to be regenerated onboard; meanwhile, some are regenerated or disposed of onshore. Somehow, the regenerated media is replaced once it is exhausted, where it can no longer absorb or adsorb the oil molecules. Having a few advantages which include low capital and operating cost as well as compact treatment unit, these two sorption techniques are suitable for vessels with less than 400 gross tonnages [16]. **Table 1** summarizes the comparison of the oil–water separation techniques.

4. Commercial OWS for bilge water treatment

Currently, many types of OWS for bilge from different manufacturers are available in the market. The separators consist of a number of separator units that

No.	Manufacturers	Name of the products	Separation techniques applied
1	Alfa Laval	PureBilge	Centrifugation
2	EnSolve Biosystems, Inc.	PetroLiminator OWS	Biological
3	Village Marine Tec.	Village Marine Tec. Oily Water Separator (VMT OWS)	Adsorption
4	Separator Spares & Equipment, LLC.	ULTRA-SEP Bilge Water Separator	Ultrafiltration (membrane)
5	Compass Water Solutions	CRP-SEP	1. Gravitational 2. Centrifugation
		ULTRA-SEP	1. Centrifugation 2. Filtration 3. Ultrafiltration (membrane)
		VG-SEP	1. Centrifugation 2. Adsorption
6	Recovery Energy, Inc.	BOSS Oily Water Separators	1. Filtration 2. Centrifugation

Table 2.
Examples of commercially available bilge water separator.

use different separation techniques. **Table 2** shows examples of companies that produced commercially available bilge water separators.

5. Conclusions

Different techniques are available and being used in the oil–water separation techniques for bilge water treatment. Typically, more than one unit of separators is needed to meet the minimum allowable discharge limit value set by the regulatory bodies. Gravitational and centrifugal methods are said to be the first technique before undergoing further polishing separation. Polishing treatment unit caters smaller droplet of oil, or emulsified oil, which cannot be treated by gravitational and centrifugal methods. To conclude, oily bilge water can be treated with proper separation techniques so that the treated bilge water discharged to the sea comply with the limit and marine pollution can be minimized and prevented. Any other treatment techniques can as well be tested and introduced to enhance the oil–water separation process in treating bilge water.

Acknowledgements

The authors would like to acknowledge the Centre for Biofuel and Biochemical Research (CBBR) and Chemical Engineering Department, Universiti Teknologi PETRONAS for the support.

Conflict of interest

The authors declare no conflict of interest.

Author details

Nurul Aini Amran[1,2]* and Siti Nor Adibah Mustapha[1,2]

1 Chemical Engineering Department, Universiti Teknologi PETRONAS, Seri Iskandar, Perak, Malaysia

2 HICOE – Center for Biofuel and Biochemical Research, Institute of Self-Sustainable Building, Department of Chemical Engineering, Universiti Teknologi PETRONAS, Seri Iskandar, Perak, Malaysia

*Address all correspondence to: nurul.amran@utp.edu.my

References

[1] Chircop A, Moreira AW, Kindred H, Gold E. Canadian Maritime Law. Canada: Irwin Law Inc; 2016

[2] Ministry of Transport Malaysia. Malaysia Shipping Master Plan 2017 to 2022: Revitalizing Shipping for a Stronger Economy. Malaysia: Maritime Institute of Malaysia; 2017. pp. 1-60. Available from: https://www.mima.gov.my/index.php/news-list/204-malaysia-shipping-master-plan-2017-2022

[3] Grote M et al. Dry bulk cargo shipping — An overlooked threat to the marine environment? Marine Pollution Bulletin. 2016;**110**(1):511-519. DOI: 10.1016/j.marpolbul.2016.05.066

[4] Al Bollmann E. World Ocean review. Physical Review E. 2010;**67**:232

[5] UNCTAD, Review of Maritime Transport. Switzerland: United Nations Conference on Trade and Development; 2014. Available from: https://unctad.org/en/Pages/Publications/Review-of-Maritime-Transport-(Series).aspx

[6] Moutzouris IC. Asset Valuation IN Bulk Shipping. City: University of London; 2017

[7] Khalid N. Malaysia's Maritime Industry: Transitioning to Industry 4.0. Selangor, Malaysia: myForesight; 2018. pp. 7-11

[8] Karakulski K, Kozlowski A, Morawski AW. Purification of oily wastewater by hybrid UF/MD. Separations Technology. 1995;**5**:197-205. DOI: 10.1016/S0043-1354(01)00083-5

[9] Office of Water, EPA. Cruise Ship Discharge Assessment Report, December 29, 2008. Washington, D.C: US Environmental Protection Agency; 2008

[10] Office of Enforcement and Complaince Assurance, EPA. A Guide for Ship Scrappers: Tips for Regulatory Compliance. Washington, D.C: US Environmental Protection Agency; 2000. pp. 4.1-4.42

[11] Magnusson K, Jalkanen JP, Johansson L, Smailys V, Telemo P, Winnes H. Risk assessment of bilge water discharges in two Baltic shipping lanes. Marine Pollution Bulletin. 2018; **126**:575-584. DOI: 10.1016/j.marpolbul.2017.09.035

[12] Tiselius P, Magnusson K. Toxicity of treated bilge water: The need for revised regulatory control. Marine Pollution Bulletin. 2017;**114**(2):860-866. DOI: 10.1016/j.marpolbul.2016.11.010

[13] Greer C, Hodson P, Li Z, King T, Lee K. Toxicity of crude oil chemically dispersed in a wave tank to embryos of Atlantic herring (*Clupea harengus*). Environmental Toxicology and Chemistry. 2012;**31**:1324-1333. DOI: 10.1002/etc.1828

[14] Wu D, Wang Z, Hollebone B, McIntosh S, King T, Hodson P. Comparative toxicity of four chemically dispersed and undispersed crude oils to rainbow trout embryos. Environmental Toxicology and Chemistry. 2012;**31**: 754-765. DOI: 10.1002/etc.1739

[15] Almeda R, Baca S, Hyatt C, Buskey EJ. Ingestion and sublethal effects of physically and chemically dispersed crude oil on marine planktonic copepods. Ecotoxicology. 2014;**23**(6):988-1003. DOI: 10.1007/s10646-014-1242-6

[16] Office of Wastewater Management, EPA. Oily Bilgewater Separators. Washington, D.C: US Environmental Protection Agency; 2011

[17] Karakulski K, Kozlowski A, Morawski AW. Purification of oily

wastewater by ultrafiltration. Separations Technology. 1995;**5**:197-205

[18] US Army Corps of Engineers, Afghanistan Engineer District AED Design Requirements: Superelevation Road Design. Afghanistan; 2009

[19] Koss L. Technology development for environmentally sound ships of the 21st century: An international perspective. Journal of Marine Science and Technology. 1996;**1**(3):127-137. DOI: 10.1007/BF02391173

[20] Kumar A. Oily Water Separator. USA: MarinerDesk; 2019. Available from: https://www.marinerdesk.com/oily-water-separator/

[21] Alfa Laval Corporate AB. PureBilge: Technical information for bilge water treatment

[22] Karhu M, Leiviskä T, Tanskanen J. Enhanced DAF in breaking up oil-in-water emulsions. Separation and Purification Technology. 2014;**122**:231-241. DOI: 10.1016/j.seppur.2013.11.007

[23] Bande RM, Prasad B, Mishra IM, Wasewar KL. Oil field effluent water treatment for safe disposal by electroflotation. Chemical Engineering Journal. 2008;**137**:503-509

[24] Watcharasing S, Kongkowit W, Chavadej S. Motor oil removal from water by continuous froth flotation using extended surfactant: Effects of air bubble parameters and surfactant concentration. Separation and Purification Technology. 2009;**70**:179-189

[25] Moosai R, Dawe RA. Gas attachment of oil droplets for gas flotation for oily wastewater cleanup. Separation and Purification Technology. 2003;**33**(3):303-314. DOI: 10.1016/S1383-5866(03)00091-1

[26] Li Z, Yang P. Review on physicochemical, chemical, and biological processes for pharmaceutical wastewater. IOP Conference Series: Earth and Environmental Sciences. 2018;**113**(1):1-6. DOI: 10.1088/1755-1315/113/1/012185

[27] Yu L, Han M, He F. A review of treating oily wastewater. Arabian Journal of Chemistry. 2017;**10**:S1913-S1922. DOI: 10.1016/j.arabjc.2013.07.020

[28] Ahmad AL, Sumathi S, Hameed BH. Coagulation of residue oil and suspended solid in palm oil mill effluent by chitosan, alum and PAC. Chemical Engineering Journal. 2006;**118**(1–2):99-105. DOI: 10.1016/j.cej.2006.02.001

[29] Metcalf and Eddy, Inc. Wastewater Engineering: Treatment and Resource Recovery, 5th Ed. New York: McGraw-Hill; 2014

[30] Zeng Y, Yang C, Zhang J, Pu W. Feasibility investigation of oily wastewater treatment by combination of zinc and PAM in coagulation/flocculation. Journal of Hazardous Materials. 2007;**147**(3):991-996. DOI: 10.1016/j.jhazmat.2007.01.129

[31] Caplan JA, Newton C, Kelemen D. Technical Report: Novel Oil/Water Separator for Treatment of Oily Bilgewaterp; 2000

[32] Oil water separator. EnSolve Biosystem Incorporation

[33] Cheryan M, Rajagopalan N. Membrane processing of oily streams. Wastewater treatment and waste reduction. Journal of Membrane Science. 1998;**151**(1):13-28. DOI: 10.1016/S0376-7388(98)00190-2